Henri Poincaré

Ferdinand Verhulst

Henri Poincaré

Impatient Genius

 Springer

Ferdinand Verhulst
Mathematisch Instituut
University of Utrecht
Budapestlaan 6
Utrecht
Netherlands

ISBN 978-1-4899-9914-6   ISBN 978-1-4614-2407-9 (eBook)
DOI 10.1007/978-1-4614-2407-9
Springer New York Heidelberg Dordrecht London

Mathematics Subject Classification (2010): 01A55, 01A70, 03-03, 11F99, 34-03, 35-03, 37-03, 70-03

Voor Claartje A.

Many who have never had occasion to learn what mathematics is, confuse it with arithmetic and consider it a dry and arid science. In reality, however, it is the science which demands the utmost imagination. It seems to me that the poet must see what others do not see, must look deeper than others look. And the mathematician must do the same thing.

—*Sonya Kovalevskaya*

Preface

It is quite an enterprise to write about a scientist who made fundamental contributions to so many fields. After working for several years on this biography, I remarked to a German colleague and friend, "I have to stop this; Poincaré is too great for me." It seemed to me that it was impossible for one mathematician to give a complete account of everything that Poincaré had accomplished. My friend replied, "That is not a good reason to give it up. Poincaré is too great for all of us." The feeling of impossibility did not disappear, but my friend's words were a kind of support, and I continued.

There was another obstruction. Poincaré was an explorer and adventurer, but of the jungles, deserts, and mountains of the spirit. He made fantastic journeys, but all those adventures took place in his mind. How can one describe such an exciting life that from the outside looks so dull? During the time I was immersed in his work, there were days that left me in such a state of excitement that I could hardly sleep. It is my hope that the reader will be equally enthralled by some of these adventures of the mind.

A number of chapters of this book have been read and criticized in manuscript. I would like to acknowledge the comments of Jan Aarts, Henk Broer, Roelof Bruggeman, Dirk van Dalen, Antonio Degasperis, Giuseppe Gaeta, Jean-Marc Ginoux, Kim Plovker, Giuseppe Pucacco, Theo Ruijgrok, and Arjen Sevenster.

Special thanks go to David Kramer, who copyedited the book and helped to sharpen the presentation through a number of queries seeking clarification.

The translations from French and German into English are my responsibility, except for Chapter 5, on the prize essay for Oscar II, where I used the beautiful book by June Barrow-Green [Barrow-Green 1997]. For Section 4.4, on Poincaré's relationship with Mittag-Leffler, the monograph [Poincaré 1999], edited by Philippe Nabonnand with many valuable notes, was a great help. Finally, the site and help of the Nancy Poincaré Archive is gratefully acknowledged.

University of Utrecht, 2012

Contents

Prologue

One sunny afternoon, on the outskirts of Nancy, in the French province of Lorraine, a woman was walking with her children, Henri and Aline. They were walking along a brook with paths on both sides, connected by several bridges. Little Henri, who was two years older than his sister, often ran on ahead with his dog, Tom. Suddenly, he noticed that his mother and sister had crossed the river and were walking on the other side. Henri's mother gestured to him that he could cross at the next bridge, but he immediately jumped into the water, which came up to his waist, and dashed across the river to rejoin his mother and sister.

This direct style of solving problems was typical of Henri Poincaré for the rest of his life. Also the impatience.

Part I
The Life of Henri Poincaré

Chapter 1
The Early Years

Jules-Henri Poincaré, called Henri, was born in 1854, in Nancy, the capital of the duchy of Lorraine, which in 1766 had become part of France. He died in 1912, in Paris, 58 years old, from a complication following an operation.

1.1 Childhood, 1854–1860

Henri's father, Léon Poincaré, was a physician with a special interest in neurology. He was a professor of medicine at the University of Nancy. His mother, Eugénie Launois, came from a well-to-do family in the provincial town of Arrancy, in the Moselle region, also in Lorraine. She was a lively and intelligent woman, and a considerate wife and caring mother of their two children, Henri and Aline (Figure 1.3). Throughout her life, she remained very close to her children. A brief Launois family tree is presented in the diagram below:

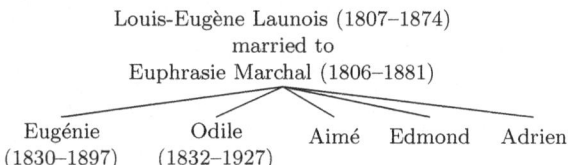

Louis-Eugène Launois (1807–1874)
married to
Euphrasie Marchal (1806–1881)

| Eugénie | Odile | Aimé | Edmond | Adrien |
| (1830–1897) | (1832–1927) | | | |

Henri's paternal grandfather, Jacques-Nicolas Poincaré, the father of Léon, came originally from Neufchâteau, in Lorraine. In 1820, he established himself in Nancy as a pharmacist, and in 1833, bought a large house called the Hôtel Martigny. This house served both as a residence and as a place of business, since part of the building had been earlier converted to a pharmacy. It is a corner house in the Rue de Guise, at that time called Rue de la Ville-Vieille, see Figure 1.1. This became the family home for Henri and Aline and their parents, as well as their grandparents and other relatives. There was also a maid, Fifine, who looked after the children, told them

F. Verhulst, *Henri Poincaré: Impatient Genius*, DOI 10.1007/978-1-4614-2407-9_1,
© Springer Science+Business Media New York 2012

Fig. 1.1 The house where Henri Poincaré was born

stories, and taught them songs. Later, when Aline married the philosopher Émile Boutroux, Fifine moved with them to Paris. Below is an abbreviated Poincaré family tree:

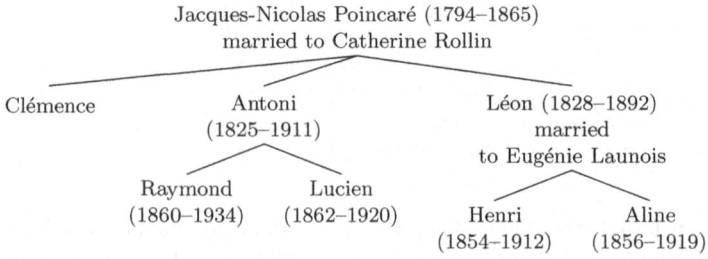

This house was in the old centre of town, not far from the ducal palace, where it stands to this day. With its many nooks and staircases, a garden, and other such places to play and hide, it was a home that stimulated the children's imagination. Grandfather Jacques-Nicolas ran the pharmacy until his death in 1865. Later, the

Fig. 1.2 Léon Poincaré and
Eugénie Launois, parents of
Henri and Aline

university of Nancy converted the Poincaré home to the Institute of Mathematics
and Physics. A plaque on the building reads as follows:

> In this house was born on April 29, 1854, Henri Poincaré. He was a member of the
> Académie Française and the Académie des Sciences. He died in Paris on July 17, 1912.

Henri and Aline grew up in what we would now call an extended family. Their
father was always busy with his patients and his research, but their mother gave them
plenty of loving attention (Léon and Eugénie Poincaré are depicted in Figure 1.2).
Their grandparents were always available for walks and story-telling. Nor was there
any lack of uncles and aunts and friends of the family who came to visit and often
stayed for several days. It was indeed a lively family.

The Poincaré and Launois Families

The Poincaré family had deep roots in Lorraine. There were Poincarés in parliament
and the military, as well as lawyers, pharmacists, scientists, and other professionals.

Fig. 1.3 Aline and Henri
Poincaré with dog Tom

There is uncertainty as to the origin of the family name: in old church and town records there appear such spellings as Poinquarrez, Poingcarré, and Pontcarré. (According to his supervisor and later colleague Gaston Darboux, Henri preferred the spelling Pontcarré.)

Henri's paternal grandfather, Jacques-Nicolas Poincaré (1794–1865), began his career as assistant pharmacist in Saint-Quentin in 1813, the year of the defeat of Napoleon at Leipzig. In 1820, he moved to Nancy, where in 1823, he married Catherine Rollin, the daughter of a locksmith. As mentioned above, in 1833 he bought the Hôtel Martigny, where he modernized the old pharmacy and used it both for preparing pharmaceuticals and as a shop. Here he lived with his wife, his children, and his sister Hélène, who was always fashionably and elegantly dressed, in contrast to the rather frugal grandmother Catherine Poincaré; Henri and Aline called her Aunt Minette.

Jacques-Nicolas had a wide range of interests. He told his grandchildren many family stories, and also taught them about the plants and animals of Lorraine. One of his achievements was the composition of a book on the flora of the Moselle valley.

Jacques-Nicolas and Catherine had three children: a daughter, Clémence, and two sons, Antoni and Léon. Clémence married a pharmacist and later moved to the nearby village of Heillecourt, four kilometres from Nancy. Henri and Aline would often walk with their mother to this village to visit their hospitable Aunt Clémence.

Antoni Poincaré (1829–1911) was a brilliant student at the École Polytechnique, where he studied civil engineering. Later, he became an inspector of state roads and bridges in Bar-le-Duc. He wrote several articles on meteorology and was acquainted with a number of professors of the École Polytechnique, for instance the prominent mathematician Laguerre. When Henri studied in Paris, Uncle Antoni visited him regularly, and following one of those occasions, Henri wrote in a letter to his parents:

> Uncle Antoni left this morning, loaded with New Year's presents for the most important people in Bar-le-Duc. On Wednesday, I had dinner with him at the Talon, and we had just been seated when a gentleman with a black beard accosted him with, "Ah, good day M. Poincaré," after which he immediately started to talk about the dialects of the Moselle region and about various activities of the Scientific Society of Bar-le-Duc. After he left, Uncle Antoni told me that this was André Theuriet, who published in the *Revue de Deux Mondes* [Poincaré 2012].

Antoni had two sons, Raymond and Lucien. Raymond Poincaré (1860–1934) played an important role in French politics, serving as president of the republic (1913–1920) during the First World War and the crucial first years thereafter. He served also for many years as minister of finance, minister of foreign affairs, and prime minister.

Lucien Poincaré (1862–1920) became general supervisor of secondary education in France (directeur de l'enseignement secondaire).

It is unsurprising, in a world in which political celebrity far exceeds scientific renown, that in Nancy, a large shopping street is named after Raymond, while a much more modest street bears the name of his cousin Henri. It is also interesting to compare the entries in the encyclopaedia *Larousse Illustré*, in which sixteen lines are devoted to Raymond, while Henri is granted only five.

Notwithstanding their important social positions, the Poincarés were by no means conformists. For example, Antoni, despite his prominent public position as inspector, refused to take the oath of loyalty to the emperor after the 1852 coup d'état.

Regarding this streak of independence, an old story circulated in the family about a relation, Gaspard-Joseph Poincaré (1762–1837), who studied in Paris and later became a Cistercian monk. In the monastery, he studied theology and mathematics but was eventually expelled for his unorthodox opinions. Outside the monastery, he still functioned as a priest until the onset of the French Revolution. He was then ordered to end his clerical activities, but he refused to do so and was imprisoned. He survived this ordeal, but later he renounced his holy orders, started a family, and became a respectable citizen.

Jacques-Nicolas wanted his son Léon to succeed him in the pharmacy, but Léon wanted to travel and had no desire for a career that would keep him behind a counter. Jacques-Nicolas made no concessions to his son, but Léon nevertheless secretly pursued a course in medicine in Strasbourg, where he successfully completed his studies. Only after his examinations did he inform his father of what he had done, adding that he now wished to continue his studies in Paris.

Jacques-Nicolas acceded to his son's request, but always maintained his allowance at a low level. Léon Poincaré became a much respected physician in Nancy, lecturing in anatomy and physiology at the medical school. At that time, around 1860–1870, there existed a number of institutes of higher education in Nancy, such as the school of forestry (École Forestière), which produced forestry engineers. There was also the widely known École de Nancy, which offered lectures in architecture, art, and industrial design. The École de Nancy played a prominent role in the Art Nouveau movement.

Following the Franco-Prussian War, the annexation of Alsace put the University of Strasbourg in German territory, and so the university was moved to Nancy, where the Nancy School of Medicine was merged with the medical faculty of the university. In 1878, Léon Poincaré obtained a chair in the medical faculty.

The Launois family represented a different side of life. Henri's maternal grandparents, Louis-Eugène (1807–1874) and Euphrasie Marchal (1806–1881), called Mémère, lived in Arrancy on land that included a large house and was part farm, part hall; today, the house is called Château Reny. It was here that holidays were usually spent when Henri and Aline were young. The trip from Nancy to Arrancy, which today can be accomplished in one or two hours, was quite a journey at that time. After travel by train to Metz, one would spend the night there in a hotel, and then in the morning take the post coach to Briey-et-Pierrepont, where the family carriage was waiting to transport them to Arrancy. Eugénie stayed there regularly with her children, her sister Odile, and numerous cousins, nieces, nephews, and other members of the family. On festival days one could find as many as sixty guests at the house, eating, drinking, walking, and playing, with grandmother Launois (Mémère) at the centre of everything. (Grandmother, it may be added, was always victorious in competitions involving numerical calculation and games of cards.)

The Launois family also boasts a number of scientists. Henri's cousin Albin Haller (1849–1925), for example, was a chemist of distinction and a member of the Académie des Sciences. Albin married Lucie, a daughter of Aunt Odile, and he and Henri later became close friends. Until he left to study in Paris, Henri would frequently visit Arrancy.

A Long Illness

During the winter of 1859, Henri, who was then five years old, became ill. His father diagnosed his illness as diphtheria, at that time a life-threatening disease for which there was no reliable cure. Henri survived, but he was left unable to walk or speak. The paralysis of his legs soon passed, but speaking remained difficult for a long time. This period of illness lasted nine months.

When the acute danger of contagion was over, Aline, three years old, could visit Henri again, but for both of them, the situation was unusual and difficult.

They found a new way of communicating in which Henri gestured and Aline learned to understand him. This gave an even greater feeling of intimacy to their relationship. This period of illness, however, made Henri physically insecure. He moved awkwardly and became shy and a bit afraid to join in rough play. Little Aline took him by the hand and guided his first steps around the house.

For some children, such a traumatic experience of a long illness followed by a period of paralysis would have serious repercussions throughout their entire lives. Yet in the case of Henri, there is no indication of such aftereffects. How, then, did he manage to remain so relatively unscathed? Perhaps in this situation of being bedridden and unable to speak, Henri's creative faculties were stimulated. In support of such a hypothesis is that fact that he made contact with his sister by gesturing. Aline, too, though only three years old, had a positive effect. During his convalescence, Henri's impatient and active nature soon took over. He wanted to persevere. In short, conditions were favourable for conquering the aftermath of his illness.

Primary and Secondary Education in France

In the nineteenth century, there were schools for primary education (*enseignement primaire*) and secondary education (*enseignement secondaire*), but until the 1880s, there was no requirement that all children attend school. For those who did, their primary schooling ended at age eleven, and that was followed by four years of secondary education, at a *collège*. One could (and can to this day) conclude one's secondary education by attendance at a *lycée* for an additional two or three years. The final examination of the lycée leads to the *baccalauréat*. Those passing this examination are then qualified to pursue higher education.

1.2 Schoolboy: 1860–1870

The brothers Antoni and Léon Poincaré had many ideas in common about education: it should have a broad scope and at the same time be substantial in content. In 1860, a friend of the family, Alphonse Hinzelin, proposed to Henri's father that the boy be tutored privately along with his daughter, who was of the same age. (Hinzelin, by the way, was one of the witnesses who signed Henri's birth certificate.) A year later, Émile Hinzelin and Aline were old enough to join them.

The lessons were not very systematic, consisting in large part of grammar and spelling, reading exercises, history, and biology. Books with pictures of flora, fauna, and geological features were used for additional instruction. Henri's relatives never got the impression that he was doing homework. It seemed that he recorded the lessons immediately in his memory.

Henri's family moved in 1862 to Rue Lafayette 6, and in October of the same year, when he was eight years old, Henri began to attend school regularly. This occurred shortly after his aunts Odile and Amélie Launois had arrived, accompanied by their sons, Henri's cousins Louis and Roger. They soon joined Henri at school, and all were excited about the novelty of new rules, noise, unknown pupils and teachers, new requirements, and homework.

Henri's class was supervised by schoolmaster Chouvenal [Bellivier 1956, p. 62]. By the end of the first week of school, Henri was first in his class in reading and first on the class honours list. This was no surprise to his private tutor, M. Hinzelin. It should be noted that the French educational system is highly competitive. Students are ranked continually, a practice that is stimulating for the excellent ones but often discouraging for those less gifted. These ranking lists are still available, and thus it is easy to follow Henri's progress at school.

The results of the first week were typical of the whole year at school. Among the 24 pupils in his class, Henri never did worse than third place in any subject or on any assignment. His private education had apparently been quite good, and of course, in retrospect, we know about Henri's intellectual brilliance. But more importantly, the rather rapid transition from the solitude and intimacy of home to the pressure and discipline of school had gone very well. In the following years, Henri maintained his position at the top of his class.

Aside from his exceptional intelligence, Henri also had an exceptional memory. His dissertation advisor and later colleague Gaston Darboux observed [Darboux 1913] that it was probably not very well known how much Henri Poincaré knew when he was young. He had only to read a book once to know its contents in their entirety; he could recall on what page and on what line of that page a specific item in the book could be found. For years following a trip abroad, he could recite all the stations at which the train had stopped and in addition, the names of all the towns and hotels where they had stayed. With such a memory, it is no surprise that he was able to master his lessons at school without making any notes.

In 1865, when he was eleven years old, Henri received, together with his cousins Louis and Roger, his First Communion, the sacrament of the Roman Catholic Church in which a child receives for the first time at the altar the Eucharist, symbolizing in a mystical way the body of Christ. According to his sister, Aline, Henri took this occasion very seriously [Boutroux 1912].

In that same year, 1865, Henri and Aline made their first big journey, travelling with their parents to the Vosges, a mountain range in eastern France, accompanied by the Xardel family. Xardel was a colleague of Léon Poincaré at the medical faculty in Nancy; his family counted five children, including a boy, Paul, the same age as Henri. Darboux [Darboux 1913], citing Paul Xardel, tells that on this occasion, the two families visited the Vallée de Ramberchamp, a place famous for its echo. Henri explained to all the theory of echoes, including the part played by the velocity of sound and the distances involved. Noting the telegraph cables along the roads, he gave a disquisition on the role of electricity in sending messages by telegraph, all in the most natural way, with no hint of conceit.

In the summer, they visited Cologne and Frankfurt, but they concluded their holidays in familiar Arrancy. Here, Henri wrote a play in five parts based on the life of Joan of Arc, to be performed by his sister, his cousins Louis and Roger, and himself. A quarrel arose between the cousins, destroying the friendly atmosphere and a number of stage properties. Henri saved the day by proposing that he transform the drama into an opera, with the actors writing their own arias. The opera was a great success in the family.

In the summer of 1867, the Poincaré family, joined by Aunt Odile and Cousin Louis, visited the World Exposition in Paris. In addition to the wonders of the exhibits, the children were amazed at the speed of service in the restaurants. It is remarkable that the children were taken along to such events, since in those days it was usual to leave the children at home; there were plenty of people to look after them. However, the Poincaré and Launois families wanted their children to have as broad and stimulating an education as possible. Both families returned to Arrancy to spend the rest of the holidays together. Inspired by their impressions of Paris, Henri, Aline, and Louis founded a political entity, the "Trinasie," a federation of three governments. There was a constitution. Henri made the appointments to the ministries, and the children invented three languages along with a common language, *Trinasien*. This federation lasted several years and was the framework for many enterprises.

During the period 1867–1868, when Henri was nearly fourteen, one of his teachers at the secondary school reported to his mother, Éugenie, "Henri will be a mathematician." When she appeared not to understand, he added, "I mean a great mathematician." However, Henri himself told Aline at the time [Boutroux 1912], "I cannot commit myself to anything. I don't know what I will do in twenty years' time." This seems a natural attitude for a boy of that age. During this same period, Henri took lessons in dancing and piano. He became an indefatigable dancer, but the piano was not a success.

The following year, when Henri turned fifteen, there was a great deal of theatrical performance, but now for a wider public. Henri's preference was for comedy, and according to Aline [Boutroux 1912], he developed a special technique:

> Henri had his own ideas about acting. He would select a member of the audience, usually an impressionable girl who laughed often and seemed to admire the performance. Between this member of the public and himself a special relation was created; she seemed lost when he was not on stage. He paid her special attention, directing cordial remarks and confidential winks to her. The result was that his acting seemed very comical, not only for the young woman but for the entire audience.

This bit of history suggests in Henri considerable social awareness and psychological insight. His troupe appeared in Nancy and in neighbouring villages, whither they would usually travel by bus and return on foot.

Henri's father, Léon, never gave up his youthful passion for travelling. In 1869, the family's annual trip took them first to the Isle of Wight and then to London. Imagine their enterprising spirit: parents with children aged 13 and 15, none of them with any English, and none having ever been at sea. But the family had prepared themselves by reading Dickens. As was their custom, they returned by way of hospitable Arrancy. Little did they know that on their next visit to Arrancy, their beloved *commune* would be devastated by war. Fortunately, Henri's grandparents' house was largely spared.

Occupation of Nancy

On July 19, 1870, France and Prussia exchanged declarations of war, and only a few weeks later, on August 14, Prussia conquered Nancy. No school diplomas were issued that summer. Cousin Louis was put on the last train to Brussels, where he continued his studies. Cousin Roger joined the resistance in Lorraine, with the risk of summary execution by the enemy were he to be apprehended. The Poincaré family considered leaving town, but rejected the idea because of the general chaos in the countryside. Léon Poincaré was placed in charge of an ambulance, with Henri, now 16 years old, a medical assistant.

A high German official was quartered in the Poincaré house, a situation that Henri immediately put to good use. Each evening, after dinner, he would converse with the man to improve his German and simultaneously learn the latest news [Boutroux 1912]. Thus he prepared himself for future travel and study while working to obtain as much information about the current situation as possible. During the occupation, news came mainly from German newspapers and other German sources. Every night, the family, using a dictionary, translated the German war news.

In November 1870, after a period of great uncertainty, messages arrived from Arrancy. The news was not good, and so mother Eugénie set out to visit her family, together with Henri and Aline. The journey was a hard one. Everywhere, they encountered soldiers; Metz was in ruins after months of fighting; desolation in the region was widespread. On their arrival, Eugénie found her parents downcast. Their youngest son, Adrien, an army officer, had been taken prisoner and transported to Prussia. There were rumours in Arrancy that the French general Bourbaki would liberate Nancy, but Henri shook his head: "That is impossible" [Boutroux 1912]. He did not expect much from Bourbaki's activities.

For the French in general and for the people of Alsace and Lorraine in particular, this was an extremely difficult period. The government was in disarray; the French army seemed powerless; large parts of France were occupied by foreign troops. The people of Nancy saw destruction, the loss of relatives, and possibly even the nearly unthinkable absorption of their town into Germany. In the end, Nancy did not suffer such a fate, but the nightmare seemed plausible enough at the time. Alsace became German, and the town of Metz was also lost. Those outcomes would poison relations between the French and the Germans for many years to come.

The annexation of Alsace in the year that followed brought many fugitives to Nancy, including a fifteen-year-old boy, Paul Appell, who became a lifelong friend of Henri.

The Franco-Prussian War

Nineteenth-century Europe saw the restoration of conservative forces everywhere. In 1852, the Second Republic collapsed and was replaced by the imperial Bonapartist regime of Louis-Napoléon (Napoleon III). Germany was still a collection of independent principalities, with Prussia the strongest state both politically and militarily. The Prussian kaiser, Wilhelm I (reigned from 1861 to 1888), was assisted in government by his "Iron Chancellor" Otto von Bismarck. Prussia put an active European policy into effect that unnerved the French and in particular made Napoleon III suspicious. In 1870, this suspicion induced France to declare war on Prussia. In those times, one still thought of war as "politics by other means," but in this case, the war became a catastrophe.

The French army was no match for the Prussian forces, and in no time, Paris was under siege. A direct consequence was the abdication of Napoleon III in 1870 and with that, the end of the monarchy. In 1871, there was a negotiated peace, but the disorder in France was enormous. After a long struggle between monarchists and republicans, the Third French Republic was established in 1875. Another important consequence was the recognition of the power and political influence of Prussia, which enabled Bismarck to push through the creation of a united Germany, in which the sovereigns of the principalities recognized Wilhelm of Prussia as their emperor.

At the peace negotiations, France was forced to cede Alsace and large parts of Lorraine, including Metz but without Nancy, to Germany. The French felt crushed. The loss of these territories would be reversed in 1918, after the First World War.

1.3 Between School and the Academy: 1871–1873

In order to be admitted to one of the elite French schools (*grandes écoles*), one must pass a national examination, the *concours*. To prepare for the exam, the usual practice was to spend a few years at a lycée, where one could pursue an elementary or more advanced course (the special lectures), leading to a bachelor's degree, the *baccalauréat*. Some of the students who followed the special lectures would take part in the *concours*.

To understand Henri Poincaré's growth and development, it is important to note that when he was young, he was considered to be a remarkably gifted lad, but not a child prodigy. He did not specialize in one subject but had a very wide range of interests, including mathematics and physics, engineering, philosophy, the arts, acting, and even writing for the stage. Many members of his immediate family and other relatives connected to these various interests offered understanding, encouragement, and security to develop his talents. In this respect, one might compare Henri with Christiaan Huygens, who in a very different era and in a different country also had very wide interests. Huygens had artistically gifted brothers, and his father was a statesman and well-known poet and composer.

During the Franco-Prussian War of 1870–1871, Henri began at the Nancy lycée to prepare for a bachelor's degree in the arts. The examination was held in August 1871. Henri's overall grade was "good." None of his grades in the individual subjects was outstanding except for Latin composition. He made, however, quite an impression with a philosophical essay titled, "How can a country elevate itself?" That was a most apropos question for France at that time.

Aline tells [Boutroux 1912] that Henri was also prepared to take the bachelor's examination in the sciences, but his teachers did not want him to take both examinations at the same time, lest it appear that he had taken the examination without adequate preparation. So Henri took the exam the following November, and it became a classic examination nightmare. He arrived a little late for the examination and in his hurry, misunderstood the first problem, which involved

geometric series. In his answer, he discussed a different problem from the one
formulated, and this yielded him a zero on the first of the two written parts. His score
on the other written question was less than spectacular, and while his scores on the
four oral examination questions were not so bad, one normally could not pass if one
had scored zero on one of the questions. The examiners, however, knew that Henri
was a brilliant student, and so they allowed him to pass with the meagre qualification
"reasonably good." Each question was graded with a score from zero to five. Henri's
grades were 0 and 2 on the written parts, and 3, 4, 2, 4 on the oral parts.

Preparation for the Concours

Henri now enrolled in the special course of study that would prepare him for the
entrance examination, the *concours*. Before he began these studies, he plunged
into the mathematics textbooks of the time. According to [Bellivier 1956], these
included *La Géométrie*, by Rouché; *L'Algèbre*, by Bertrand; *Cours d'Analyse de
l'École Polytechnique*, by Duhamel; and *La Géométrie Supérieure*, by Chasles.
The last two books on the list are particularly remarkable. Jean Duhamel's calculus
book was a textbook for the École Polytechnique, where Duhamel (1797–1872)
himself lectured. Michel Chasles's geometry text emphasized the complementary
roles of analysis and geometry; it was original, difficult, questioned by a number of
colleagues, yet written in an engaging style. Chasles (1793–1880) formulated his
approach in his 1846 inaugural lecture for the geometry chair at the Sorbonne as
follows (see also [Chasles 1880]):

> If one knows that the subject of geometry is the measure and characteristics of space, then
> one knows how extended this field is, and one does not even know where the boundaries
> of this domain end. For space that one imagines changes shape infinitely often, and the
> features of each of the forms arising in nature or those that the human spirit can imagine are
> themselves extremely numerous, one can even say inexhaustible.

That in itself is exciting enough, but Chasles continued by proposing a drastic
change in the way geometry was practiced. As understood and practiced at the
beginning of the nineteenth century, for instance under the influence of Joseph-
Louis Lagrange, classical geometry had become very analytic; in short, it now
emphasized formulas over pictures, analysis over synthesis. Lagrange himself, in
the introduction to his *Mécanique Analytique*, writes:

> One will not find figures in this work. The methods that I explain herein require neither
> geometric nor mechanical constructions or arguments, but only algebraic operations forced
> by regular and uniform steps.

Chasles, in contrast, has this to say:

> One can see the respective advantages of Analysis and Geometry: the former leads, with the
> miraculous mechanism of its transformations, quickly from the starting point to the point

to be reached, but often without revealing the road that was travelled or the significance of the numerous formulas that have been used. Geometry, on the other hand, derives its inspiration from thoughtful consideration of things and from the ordered arrangement of ideas. It is obliged to discover in a natural way the statements that Analysis could neglect and ignore.

When Chasles's *La Géométrie Supérieure* appeared in 1852, his vision of geometry, using, of course, analysis as a support, was rather revolutionary. It seems likely that it became an important influence on Henri's way of thinking, especially considering the fruitful combination of analysis and geometry that is typical of his methods, for instance in the quantitative and qualitative theory of dynamical systems that he would later develop. This supposition is supported in [Poincaré 1905b, Chapter 5, p. 153], where Poincaré writes about the analytic and geometric images evoked by the Laplace equation:

Thanks to these images, one can see at a glance what pure deduction will show only after successive steps.

During the lycée year 1871–1872, one of his teachers persuaded Henri to sit as well for the entrance examination of the forestry school (L'École Forestière). He finished second on the forestry school examination, while at the end of the first year, the general examination at his own school, the lycée, brought him a first place. In retrospect, from today's vantage point, the advice that Henri sit for the examination at the forestry school seems strange, but one must realize that the forestry school in Nancy was a prestigious engineering school. It had been founded in 1824 and offered a varied curriculum, including, of course, mathematics. It enrolled many foreign students. The mathematician Émile-Emmanuel Regneault (1834–1866), an active researcher, was a member of the staff.

However, the most exciting event of 1872 occurred almost accidentally. In early summer of that year, the mathematical problem that had been posed at the concours of the École Polytechnique was published. Its solution would be published only later. During this time, Henri's mother noticed that her son was wandering about the house at all hours. In fact, he was trying to solve the problem. And he succeeded! This feat became known outside the family. Paul Appell, who began the special course at the lycée in September 1872, tells:

On one of the first days of the school year, a pupil pointed out Henri Poincaré to me and said, "There is a clever fellow. He came in second at the forestry school, and he took first place in elementary mathematics at the general examination, and all by himself he solved the problem posed by the École Polytechnique for this year." Poincaré's outward appearance struck me; at first glance, he did not seem the usual sort of clever pupil. He seemed preoccupied with his thoughts; his eyes were somehow veiled with thought. When he spoke, his eyes revealed his good nature; they were eyes at once twinkling and profound. I felt attracted to him, and since we were both day students, we exchanged some words on leaving school. I was struck by his manner of speech, jerky and interrupted by long silences [Appell 1925a].

His mode of expressing himself represented a potential impediment to his ambitions. Appell notes:

> From the first time you heard him in class, his high level of intellect was clear. When he answered questions, he left out all the reasoning in between, and then he was so terse and concise that the teacher always asked him to clarify his answer, adding, "If you answer like that at the examination, you will risk not being understood" [Appell 1925a].

The mathematics teacher who made this observation was Victor Elliot (1847–1894).

Like most teachers of the special classes, he was an expert in his field. He had been at the École Normale Supérieure and would defend his dissertation in 1876 under the direction of Puiseux, Hermite, and Briot, all of them outstanding mathematicians. In the latter part of his career, he chaired the faculty of the University of Besançon.

It was not only Henri's terseness that people found irritating at first. The teacher Elliot and the other pupils noticed that Henri did not write down much. He always had but a single sheet of paper on which he made the occasional note. And not only that; after several days, they observed that it was always the same sheet of paper. Was he really a serious student? That certainly turned out to be the case. He knew what he was about, and moreover, was always cheerfully ready to assist his peers. Their irritation soon changed to admiration and friendship.

The meeting between Henri and Paul Appell was the start of a lifelong friendship. Appell describes [Appell 1925a] how after school, together with another friend, Hartmann—like Paul, from Alsace—they used to take long walks, during which the three friends talked about mathematical problems, and also about philosophical questions, war, the German occupation, and the political future of France. The emperor had abdicated, and they were all for the creation of the Third Republic. According to Appell, nearly all the pupils at the lycée supported the republic.

Henri and Paul were looking toward future studies at the École Polytechnique. Their teacher Elliot, however, advised them to apply as well to the École Normale Supérieure, with the result that they had to prepare for two entrance examinations in the summer of 1873.

The Elite French Schools (Les Grandes Écoles de France)

The *grandes écoles* in France represent a type of higher education that is outside the general structure of the French university system. They were founded in order to open up higher education to a broader class of students based on merit. With the right type of bachelor's degree, one has access to a university. For the grandes écoles, in contrast, one must take an entrance examination, for which one prepares in the special classes at a lycée; before enrolling in such a special class, one may obtain a bachelor's degree in the arts or the sciences.

The grandes écoles were founded around 1800, and many of France's greatest political and academic luminaries graduated from these schools. In Henri's time, they included the following:

- The École Normale Supérieure, founded by Monge and Carnot.
- The École Polytechnique, an establishment with a military character and a military general as director. The students wear uniforms and on graduation generally become regular officers or reservists.

- The École des Mines, an establishment for mining engineers with emphasis on science and engineering.
- The École Spéciale Militaire de Saint-Cyr, which provides the standard education for regular army officers.

The lecturers of these four establishments were eminent mathematicians, scientists, and engineers. Today, there are many more grandes écoles, for instance those established by the École Normale Supérieure outside Paris as well as new establishments that emphasize government, management, and economics.

Admission to the Grandes Écoles: The Concours

In 1883, Henri Poincaré and Paul Appell were the only candidates from the special classes in Nancy for the examination of the Grandes Écoles. The examination (*concours*) for admission to the École Normale Supérieure took place in July in Paris, with Darboux as one of the examiners. As happened before in his examinations, things did not run smoothly for Henri. The examination topic was the projection and intersection of two quadrics, and Henri did not take the obvious classical approach. Moreover, he made an error in drawing the figure. After the oral examination, one of the examiners concluded that he had expressed himself badly. Henri received the ranking five, while Paul came in third.

Months later, this result came up for discussion at the École Normale Supérieure. Through Paul's mediation, the vice-principal, Bertin-Morot, even arranged a meeting with Henri to discuss the unfortunate outcome of the examination. But by then it was too late to change the course of events.

The written examination for the École Polytechnique took place on August 4–6, 1873, in Nancy. Henri was nervous, and at the conclusion of the exam expressed his disappointment with his performance. The exam problem was geometric, with the following formulation:

> Given is a quadratic surface S with two points, A and B, on the surface. There exist infinitely many quadratic surfaces R that are tangent to S at A and B. Find:
>
> 1. the locus of the centres of the surfaces S;
> 2. the locus of the contact points of these surfaces with the tangent planes that are parallel to a given plane;
> 3. the locus of the contact points of the same surfaces with the tangent planes passing through a given straight line.

On leaving the examination hall, they found a world transformed. All Nancy was bursting with excitement; flags hung from every flagpole, and throngs of people were celebrating everywhere. On that day, the German troops had left the city. Paul had mixed feelings, for his native region, Alsace, would remain a part of German territory.

The mathematicians on the committee for entrance to the École Polytechnique were Briot, Bouquet, and Puiseux. As to the outcome, there was nothing to worry about, for in fact, Henri's examination results were very good. Later, at the oral examination in Paris, he made a deep impression on the examiners. An acquaintance from Nancy, De Roche du Teilloy, was present [Bellivier 1956]. He recalled that Henri was asked to give a proof from elementary geometry. Without hesitation, he provided a correct answer, but the examiner then asked for a more elementary proof. Henri's immediate answer to this was also correct, but it contained trigonometric arguments. The examiner now asked for a proof containing elementary geometric arguments only and was again answered at once, now to his complete satisfaction. The examiners were delighted. Henri received first place on the École Polytechnique examination, and it is there that he would continue his studies. Paul enrolled at the École Normale Supérieure. In the future, now at separate schools in Paris, they would have to make more of an effort to meet. The period of daily contact was over, at least for a number of years.

Aline knew that her brother would soon move out of the house. This prospect made her pale and thin, to such an extent that in the summer of 1873, her father took the whole family to the seaside for several weeks; as usual, the last part of the holidays was spent in Arrancy, damaged as it was. But it remained a difficult time for Aline.

Chapter 2
Academic Education: 1873–1879

In the fall of 1873, Henri, then 19 years old, travelled to Paris, accompanied by his mother and sister, to enroll in the École Polytechnique. While in Paris, Henri's mother and Aline stayed with the Rinck family, old friends from Lorraine, whose son, Élie Rinck, was of the same age as Henri. They remained in Paris a week, during which time they visited Henri and saw him for the first time in uniform (see Figure 2.1); they found it very difficult to say goodbye. The feeling was mutual, attested by the fact that during his first two years in Paris, Henri wrote hundreds of letters home, more than two a week.

2.1 A Difficult Year

As we shall see, life at a military school was not easy. In 1873, there were more than two hundred freshmen (indeed men; women were not admitted), and all of them had to attend lectures, perform military drill and learn to use weapons, stand guard, and take turns at orderly duty. In the relatively few free evenings, Henri visited the hospitable Rinck family or his relatives the Olleris, or else went to the theatre. Socially, his relations with his classmates were more or less as they had been in Nancy. He was much too absorbed in his own thoughts to have an active social life. During breaks, he often walked alone, but he was helpful when it was necessary, and he was respected because of his knowledge and insight. He followed the lectures with his arms crossed and often without taking notes. At times, he would suddenly become socially involved, and this, too, was accepted by his classmates. We will discuss some of these occasions later on. Appell [Appell 1925a] describes a typical discussion of a problem between Henri and a classmate:

> If someone asked him a question, he took his arm confidentially and forced him to walk with him but in a very irregular way. Suddenly he plunged forward, halted all at once, and then walked backward. It was a little bit analogous to what the physicists call Brownian motion.

F. Verhulst, *Henri Poincaré: Impatient Genius*, DOI 10.1007/978-1-4614-2407-9_2, © Springer Science+Business Media New York 2012

Fig. 2.1 A slightly
bewildered-looking Henri
Poincaré in the uniform of the
École Polytechnique

Henri found some of the lectures interesting, in particular those of Hermite in
analysis. In that year, his other professors were Résal (mechanics), Mannheim
(geometry), Faye (astronomy), Cornu (physics), Frémy (chemistry), and Zeller
(history and literature). Outside of lectures, the students were supervised in their
studies by Bonnet and other lecturers, including Halphen. Pierre-Ossian Bonnet
(1819–1892), see Figure 2.2, formulated and proved a number of important results
in differential geometry. The Gauss–Bonnet theorem uses the notion of geometric
curvature and is a well-known example of his achievements (Gauss formulated the
theorem in terms of an example; Bonnet supplied the general theorem). George
Henri Halphen (1844–1889) served as an officer in the French army during the
Franco-Prussian war of 1870–1871, for which he was decorated. His mathematical
career began after the war. His research was concerned with invariants of differential
equations and differential geometry. He died at age 44.

In his letters home, Henri had little to say about his scholarly achievements. The
following fragment from a letter is an exception:

One day Hermite was ill; Laguerre, who replaced him, discussed a certain problem during
his lecture. Because the writing on the blackboard was not clear, I had made no notes. I paid
no attention to it, but a few days later, a classmate asked me whether I could explain the
problem to him. I answered that I had made no notes but that I would reconstruct Laguerre's
proof. This I did, or I thought that I did, but meanwhile, I felt somewhat uncertain, for I had
no use in the proof for the only remark that I had written down. That evening, the student
was called up for a preliminary examination, and Halphen asks him exactly this problem.
The student presents my proof. The examiner asks whether it is his own. The student looks
me up, asks whether it is mine, and returns then to Halphen to tell him. Halphen says that he
is not surprised. Halphen informs Laguerre, who sends for me and tells me that my proof is
simpler than his own. It will replace his proof in the publications that, I believe, are adorned
by the pompous name of *Archives of the École Polytechnique*.

Fig. 2.2 Pierre-Ossian Bonnet, director of studies at the École Polytechnique when Henri Poincaré was a student

At the beginning of 1873, the students were ranked as follows: 1. Poincaré, 2. Bonnefoy, 3. Petitdidier. The same three students continued as the highest ranked in their class until the final examination. In the spring, however, Henri's mood deteriorated, and he dropped from first place. He found the level of instruction too low, and he missed the challenge of difficult problems. In May 1874, he wrote dejectedly:

> Here it is like a gigantic machine whose motion one has to follow on pain of being overrun; people do what twenty generations of the École before us did and what $2n + 1$ generations of recruits after us will do.
>
> One needs here only two aspects of one's intelligence: memory and eloquence. To understand a course you have only to work, and that is why everybody can overtake me if he really wants to by grinding away. The examiners never ask anything exciting.

He was not always so dejected, quite the contrary, but he continually saw his two years at the École Polytechnique as a necessary phase on the road to his future instead of as an experience that was interesting in itself. Would he have been better placed at the École Normale? Perhaps. When visiting his friend Paul Appell at the École Normale, he regularly met some of the school's lecturers. For example, he once dined with the mathematician Briot. Those meetings gave him a great deal of pleasure. The École Polytechnique did not lack brilliant and stimulating lecturers, such as Hermite, Laguerre, and Halphen. However, they were not very prominent in a teaching environment geared to the education of officers and engineers.

Nonetheless, Henri's creativity in his first year could not be stopped. In October 1874, his first research paper appeared, and of course its topic was geometry: "Démonstration nouvelle des propriétés de l'indicatrice d'une surface" appeared in *Annales de Mathématiques*, 2e série, vol. XIII. It is an interesting exercise in curvature and osculating surfaces. It certainly should have made an impression that a student had written such a paper on his own and had it published, but what did the professor of geometry think about it? That will become clear indirectly.

We look first at an incident that occurred early in 1874. For each class meeting, a student was appointed to summarize the preceding lecture. Preparation for this was a joint effort, with the best students helping out. One day, the students' preparations were not going well, and they appealed to Henri. He said, "stop looking; the problem is wrong." Since their summary was an officially assigned task, the students informed the principal for teaching, Ossian Bonnet. The lecturer concerned, professor Mannheim, was asked for an explanation. Mannheim insisted that the problem formulation was correct, but much later, the problem associated with that lecture was removed from the curriculum.

At the end of Henri's first year, the geometry examination was supervised by Jules de la Gournerie, called Gournard, who was a friend of Mannheim. Henri received poor marks for geometry, and he came in second on the examination overall, after Bonnefoy but ahead of Petitdidier. It was clear that Gournard had penalized Henri severely for the quality of his drawings: drawing had always been his weak spot. The whole school was in turmoil over this injustice. Madame Rinck, who always welcomed Henri warmly during his student years, wanted to lodge a formal complaint, but Henri discouraged her. It was enough that those in positions of importance, such as the principal Bonnet, shared in the general indignation.

Finally, the summer holidays of 1874 began. As Aline noted, "We lived as if at a permanent party. We undertook everything Henri liked, everything that gave him pleasure" [Boutroux 1912].

2.2   Second Year at the École Polytechnique

The second year started well, with the publication of Henri's first scientific paper, but Henri's mood was not always cheerful. He suffered from a lack of intellectual stimulation and the feeling that the fight for first place in his class did not suit him. The struggle to be first in one's class was more than a competition for superiority over one's classmates; in France, achieving first place in a school ranking opens the next step in one's career. Henri, however, was interested in matters far removed from such trivial school affairs. Fortunately, he got on well with his classmates, and furthermore, it helped that in the second year, he could welcome some of his former comrades from Nancy who arrived at the École Polytechnique, among them Hartmann. He got on well with Bonnefoy, his most serious competitor in the rankings.

His many letters home continued, including letters written separately to Aline, but they became more carelessly written and contained less and less information about his daily life in Paris. On several occasions he was admitted to the infirmary of the École for a few days, but his illnesses were not serious. The worst of it was that the topic he enjoyed most, geometry, was in the hands of Mannheim and Gournard.

In 1875 came the examination at the end of the second year. Henri lost points in topography, drawing, and architecture. The class ranking became 1. Bonnefoy, 2. Poincaré, 3. Petitdidier.

The talented students Bonnefoy and Petididier would die young in mining accidents. Henri's difficulties with Mannheim had a sequel—or better, a settling of accounts, one might say. Following the death of Laguerre in 1886, a new member of the Académie des Sciences was elected the next year. Both Poincaré and Mannheim were on the list of nominees. Poincaré, 32 years old, was elected with 34 votes, while Mannheim, who could look back on a long and productive career in geometry, received 24 votes.

2.3   L'École des Mines

In November 1875, Henri, who was then 21, continued his higher education at the École des Mines in Paris, together with his École Polytechnique classmates Bonnefoy and Petitdidier. Today, in the twenty-first century, with practically unlimited educational possibilities, this choice seems a rather unlikely one. But there was not so much choice in 1875. The École des Mines was (and is) an excellent engineering school that provides an education for a useful and socially important profession: mining engineer. Another possibility for Henri would have been to attend the University of Paris (the Sorbonne), but that would have given him a more general education, not one that would directly qualify him for a profession. Henri would finish the mining school successfully, but his interest was in a few topics only, particularly mineralogy. With no particular effort, he remained one of the best students, but he lost his interest in the rankings.

Becoming a Mining Engineer

In his first year at the École des Mines, Henri's courses included mine management and machinery, metallurgy, mineralogy, geology, palaeontology, assaying, drawing, and English.

The director, Daubrée, who was a distant relative of Henri, maintained that Henri should not pay attention to mathematics while at the École des Mines. In December, Henri wrote home:

> This morning I have been to see M. Daubrée. He was very nice and told me, as had Uncle Antoni, that he advised me not to do mathematics before I had finished school. It appears that Bonnet had asked that I be given dispensation for certain hours so that I could attend lectures at the Sorbonne. He told me he had refused this. I answered him that I understood perfectly.

The point was that Henri did not need the lectures at the Sorbonne to prepare for their mathematics examination. Bonnet provided him with exercises, and in August 1876, he passed the Sorbonne's examination in mathematics.

During this first year, he visited Bonnet regularly at the École Polytechnique. Thus at the beginning of 1876, Henri wrote, "On Wednesday I was again visiting Bonnet, who was very nice. Also I saw Bouquet, who was as nice as was possible for him. He lent me an old book that I needed."

Studying at the École des Mines with parallel study at the Sorbonne did not interfere with Henri's social life. He kept up his stream of letters home, although at a pace slightly slower than that of his first two years in Paris. The letters describe visits to the theatre, visits to relatives, and political questions, and often they contain humorous verse. After the experiences of the Franco-Prussian War and the annexation of Alsace, there remained considerable apprehension in Nancy about the military might of Germany and in particular about the question of the possible annexation of Lorraine, Nancy included. Henri wrote to his family that such a turn of events seemed highly improbable to him, since it would require a permanent military occupation of the hostile French population of Lorraine. He added this:

> What seems more probable is the Prussian annexation of Belgium and Holland. This would be very unfortunate for us, for it would double the length of the border with Germany, and it would double the German navy; it would present Germany with rich colonies, not to mention the industrial richness of Holland and the abominable military position that would arise for us.

Henri went on to suggest that after Belgium and Holland, Bohemia and the countries to the east of Germany would be next in line. Roughly sixty years later, all of Henri's fears were realized.

Now and then, the name of Émile Boutroux appeared in letters from home. In 1876, Boutroux was appointed to the university of Nancy to lecture in philosophy. He became a regular visitor to the Poincaré family. Émile Boutroux married Aline in October 1878.

The young philosopher was interested in graphology, which led to frequent discussions with Henri's cousin Raymond Poincaré, who was studying philosophy in Nancy, and with Henri in Paris. In the fall of 1877, Raymond moved to Paris to study law. He took a room adjacent to Henri's in the Boulevard St. Michel. The philosophical discussions between the cousins that had been carried on at a distance now continued much more intensively.

As mentioned above, graphology, the study of the relationship between character and handwriting, interested Henri. In a letter home, he expressed surprise that women, who in his experience often think in a haphazard way and without much logic, often have neat and well-ordered handwriting. This caused him to reflect on his own writing:

> As far as I am concerned, I find that my main characteristics can be recognized in my handwriting. The way I make the last letter of each word very small illustrates that I am bad at waiting. My pliability shows in the softness of rounding off, corrected only by the first influence. Consider my n's and my u's, which look like Greek ω's and not like the German w's as in your case. Consider, on the other hand, the way my lines are positioned

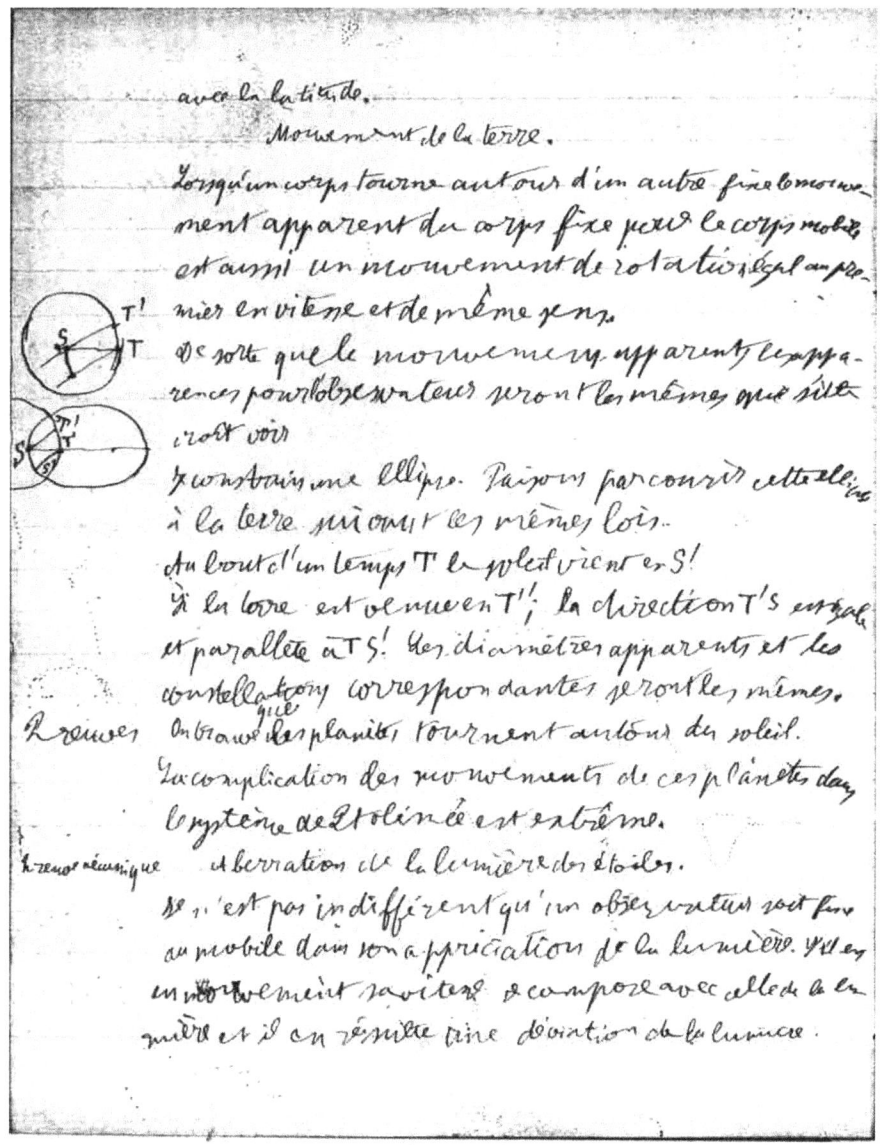

Fig. 2.3 Henri Poincaré on the Earth's motion

like vehicles emerging fresh from their village (what a difference compared to the Prussian coordination of the lines of Barrois). It is this particular thing that gives away my total absence of the bureaucratic feelings that are so widespread among the French people.

(For Barrois, read Raymond Poincaré.) For a sample of Henri's handwriting, see, for instance, the page depicted in Figure 2.3.

Fig. 2.4 Cousin Raymond
Poincaré (1860–1934) shown
in his position as a leading
politician. He became prime
minister and president of the
republic

Raymond Poincaré (1860–1934)

Raymond Poincaré, see Figure 2.4, was one of the sons of Henri's uncle Antoni. He was
born in Bar-le-Duc and studied law at the University of Paris. As a lawyer, he would later
defend Jules Verne against a libel suit. He became a government minister in 1893 at the
age of 33, and was prime minister during five government periods. Raymond was president
of the republic in the critical period 1913–1920, which covered the First World War and
the post-war treaty negotiations (Treaty of Versailles). He was a hardliner regarding the
relationship with Germany and the exaction of war reparations.

His brother Lucien Poincaré (1862–1920) was a physicist who became inspector-general
of public instruction. Raymond and Henri always stayed in contact, even in their later
careers, discussing, for instance, appointments and the awarding of medals to distinguished
people. Both of them were members of the prestigious Académie Française.

Completion of Studies at the École des Mines

The École des Mines organized excursions of several days for its students, but as
a rule, there were two longer stays in a foreign country during the course of study.
In the summer of 1877, at the end of the second year, Henri travelled, together
with his classmate Lecornu, to Austria and Hungary. Following that trip, he wrote
two reports, one on the coal mines of Hungary and the other on the pewter industry.
Lecornu later wrote to friends how cheerful Henri was during their trip and with how
much pleasure he received and read the long letters that Aline wrote to him [Appell
1925a].

A similar educational tour was made at the end of the third year, in 1878, this time together with Bonnefoy. They travelled to Sweden and Norway, and again following this trip, Henri wrote two reports on mining operations.

In June 1878, Henri's studies at the École des Mines came to an end. The final ranking had Henri third, after Bonnefoy and Petitdidier, but the result did not seem to interest him. The following year, in March 1879, Henri Poincaré, 24 years old, was appointed to the post of mining engineer in Vesoul, relatively close to Nancy. The appointment was made by the National Inspection of Mines, with the formulation "appointment to ordinary mining engineer of the third class charged with the mineralogical subdistrict Vesoul and in addition the supervision of the railways in the east." As we will see, the job was not without danger. Bonnefoy and Petitdidier also became mining engineers. They died in their late twenties from accidents that occurred in the course of their duties.

Henri's activities in Vesoul were short-lived. In December 1879, he was appointed to a lectureship of mathematics at the Faculté des Sciences of Caen, in Normandy. Formally, he remained his whole life a member of the corps of mining engineers. On June 16, 1910, he was appointed inspector-general of mines, in this case most likely an honorary title.

2.4 Dissertation in Mathematics

During all the activities of these years, Henri's mathematical discussions and research had never been interrupted. It seems that Henri almost casually wrote his dissertation during his second and third years at the École des Mines; for mathematical details, see Section 9.1. Its inspiration was from a paper by Briot and Bouquet in the *Journal de l'École Polytechniqe* [Briot and Bouquet 1856] dealing with solutions of differential equations. As a first result, Henri wrote a short paper, which he submitted to that same journal [Poincaré 1878]. The dissertation, titled *Les propriétés des functions définies par des équations aux dérivées partielles* [Poincaré 1916, Vol. 1], was submitted at the turn of the year 1877–1878 to his supervisors Darboux, Laguerre, and Bonnet. It took some effort for Henri to get their comments. In 1878 he wrote:

Darboux resides at number 36
In the same house as the good cousin.
His advice I received with great pleasure
And a short time after that a long sermon
Filling ten large pages;
Off to Laguerre where I was not so lucky,
I wanted to be counselled, but alas,
I found the door closed, and infuriated,
I headed for Ossian, and there a wooden door as well.
But I will find him some day, thank God.

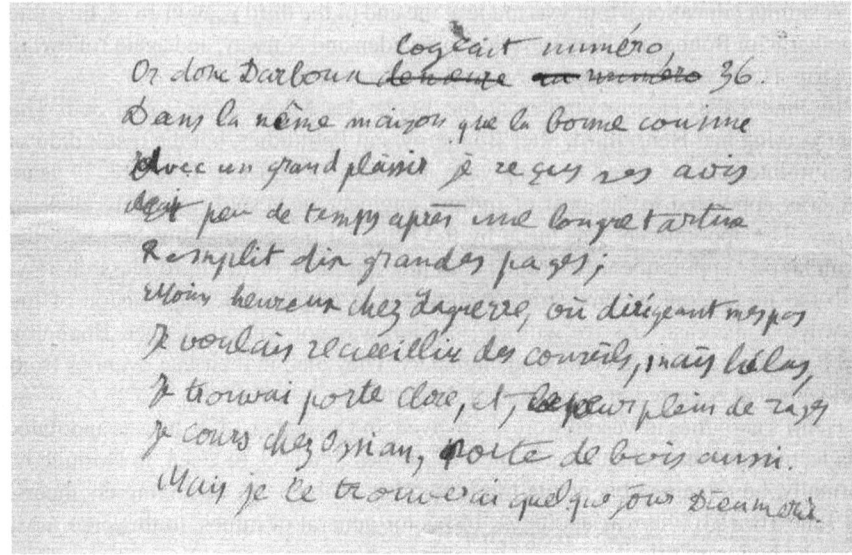

Fig. 2.5 The original text of Poincaré's poetic discontent (courtesy Archive Poincaré Nancy)

The "good cousin" was the daughter of his relatives the Olleris. The French original is reproduced in Figure 2.5; the text reads as follows:

> Or donc Darboux logeait numéro 36
> Dans la même maison que la bonne cousine.
> Avec un grand plaisir je reçus ses avis
> et peu de temps après une longue tartine.
> Remplit dix grand pages;
> Moins heureux chez Laguerre où dirigeant mes pas
> Je voulais recueillir des conseils, mais hélas,
> Je trouvais porte close et, le coeur plein de rage,
> Je cours chez Ossian, porte de bois aussi.
> Mais je le trouverai quelque jour, Dieu merci.

The dissertation was accepted on August 1, 1879, one and a half years later. It had been written hurriedly, and the supervisors made many critical remarks. Much later, Gaston Darboux wrote this about his role as supervisor [Poincaré 1916, Vol. 2]:

> Joseph Bertrand used to say that the article in which Briot and Bouquet explained their results had brought the greatest advance in this part of analysis since Euler. Henri Poincaré made his first appearance by studying and perfecting that great work. In the dissertation that he submitted in 1878, he threw himself at a still more difficult question, the integration of partial differential equations with an arbitrary number of independent variables. At first glance, it was clear to me that the manuscript went beyond the usual and contained enough material for several good dissertations. But to give a precise idea of the way Poincaré worked, we must not shrink from stating that there were many points begging for correction or explanation. Poincaré thought intuitively.... It was easy for him to make the corrections

and cleaning up that I found necessary. But he explained to me later that at the time I asked him for it, he was occupied with completely different concepts. Whatever the case, his dissertation is valuable because of a number of new and important ideas.

The new ideas to which Poincaré was referring in his later conversation with Darboux are in large part contained in his revolutionary memoir [Poincaré 1881] and his articles on quadratic and cubic forms and their invariants and the so-called Fuchsian functions, published in 1880 and 1881 (and still later); see [Poincaré 1916]. The Fuchsian, or automorphic, functions are discussed in the next chapter and in Chapter 8. The memoir is discussed in Section 9.2. It presents a completely new approach to the theory of nonlinear second-order differential equations. It gives a classification of singular points, the index theorem for closed curves, the idea of "consequents," or the Poincaré map for plane systems, and the basic ideas of the Poincaré–Bendixson theorem for limit cycles. It is now part of the general theory of ordinary differential equations, a topic on which Poincaré would publish a great deal in the years to follow. The dissertation can therefore not be separated from the memoir [Poincaré 1881] on the *global* qualitative and quantitative analysis of differential equations in the plane.

In the dissertation itself, the treatment is local, with first-order partial differential equations analysed with characteristic equations that may contain weak singularities. This leads to a technically complicated analysis with many different cases. It is understandable that the supervisors needed time to digest the material and also that they asked for examples to illustrate the theory. Unfortunately, there are not many examples presented. Of great interest in the thesis are the new concepts introduced by Poincaré. We mention the algebroid functions, the concept of what is today called a Poincaré domain, and the concept of resonance of eigenvalues. The last two ideas will return often in Poincaré's work on dynamical systems, for instance for the equations of the solar system and even more so in general approximation methods using normal forms, the so-called Poincaré–Dulac normalization.

Chapter 3
Impressive Results in Vesoul and Caen

In examining the lives of creative people, including scientists and artists, one frequently observes an initial period of acquisition of knowledge and practical skills followed by a burst of activity with occasional interruptions. For Henri Poincaré, this watershed came around 1878. In his case, however, the enormous flow of significant results continued uninterrupted throughout his life.

3.1 Mining Engineer in Vesoul

The period from the summer of 1878 until December 1879 was a busy and disquieted time for Henri. He had to complete his studies at the École des Mines, finish and defend his mathematical dissertation, and also find a job. On March 28, 1879, he graduated as a mining engineer and obtained an appointment as inspector of mines at Vesoul, in eastern France. He had requested an appointment that would put him not too far from Nancy.

He arrived in Vesoul on April 3, 1879, nearly 25 years old, with his main responsibility the mines of Ronchamp, about 30 kilometres from Vesoul. As a mining engineer, Poincaré's job was to assess mines for their production capacity and safety, including structural integrity, ventilation, and the localization and removal of inflammable gases.

During the first few months, the work of descending into mines and writing reports was routine. However, in the newly opened Magny mine (see Figure 3.1), an explosion occurred early in the morning of September 1, 1879. Of the 22 men on shift, 16 were killed by the explosion at a depth of 650 metres. While the rescue operation was in progress, Henri Poincaré descended into the mine to begin his investigation. In an extensive report, he described the ventilation and gallery system of the mine, suggesting as a possible cause of the explosion a perforated safety lamp that had been found (the light necessary for the work of breaking the coal loose from the coal seam was provided by safety lamps; there was of course no electricity).

F. Verhulst, *Henri Poincaré: Impatient Genius*, DOI 10.1007/978-1-4614-2407-9_3,
© Springer Science+Business Media New York 2012

La Haute-Saône Pittoresque - c. L. B

2109 - RONCHAMP — Puits du Magny

Fig. 3.1 The Magny mine in Ronchamp, near Vesoul, where an explosion took place in 1879. Henri Poincaré descended into this mine and submitted a report

His last report as an inspector in Vesoul was on a new ventilation system for the Magny mine, which he wrote with his colleague Trautmann. It was signed and presented on November 30, 1879 (see Figure 3.2), a day before he took up a new position in Caen as a lecturer in mathematics. He continued to hold a position in the Corps des Mines, the organization of mining engineers. In 1893 he was promoted to chief engineer, and in 1910, as mentioned above, he became inspector-general of the French mines.

In Vesoul, Henri divided his time between his work as a mining engineer and mathematics. He was qualified in both, and he practiced both professions. In that period, his cousin Raymond Poincaré wrote two novels. In Vesoul, Henri also wrote a novel, a romantic story that never was published. The manuscript has been lost, but a summary of the plot can be found in [Bellivier 1956].

3.2 Lecturer in Caen

On December 1, 1879, Henri Poincaré was appointed as a lecturer in mathematical analysis at the University of Caen, see Figure 3.3. Around the same time, Paul Appell obtained a university position in Dijon, and Émile Picard took up a post in Toulouse. According to Darboux [Darboux 1913], the French university authorities made an effort to appoint scientists of superior quality to the universities in the

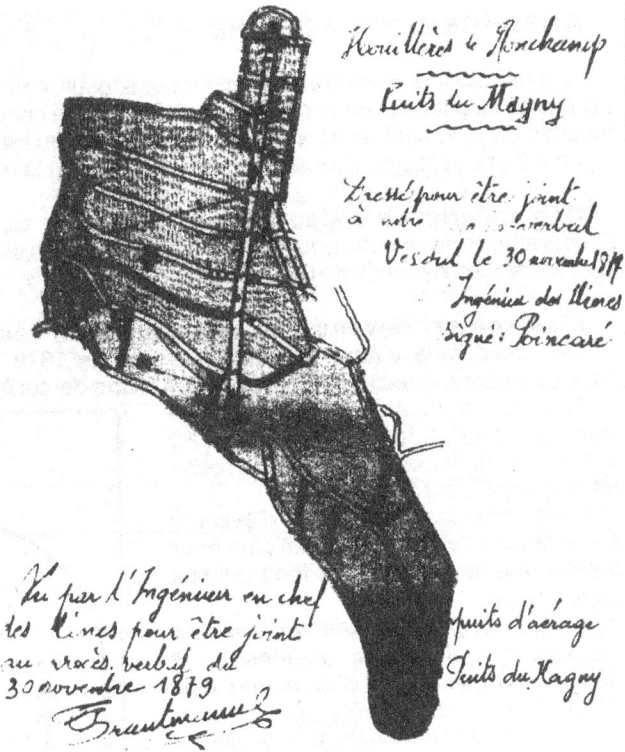

Fig. 3.2 The front page of the report for a new ventilation system of the Magny mine in Ronchamp, signed by Poincaré and Trautmann (courtesy Musée de Mines, Ronchamp)

provinces rather than have all of them concentrated in the capital. A nice idea, to be sure, but two years later, these three were together again in Paris.

For Henri, it was a time of many changes, above all a new position in a new city, with time now to think more intensively about new mathematical problems. Observing the marriages of relatives (including his sister, Aline) and friends stimulated his own interest in finding a partner. Around 1880, Henri made the acquaintance of Louise Poulain d'Andecy. They married on April 20, 1881, a few months before he was appointed to a position in Paris at the Sorbonne.

In Caen, he renewed his acquaintance with his former classmate and travel companion Lecornu, who was the local mining engineer. Lecornu recalled [Bellivier 1956] how they spent New Year's Eve in 1879. Henri seemed to have been even more preoccupied than usual:

> I remember that I asked him to come over for dinner on December 31, 1879, along with my parents. He spent the evening walking up and down, did not listen to what was said, and answered with only a few words. He lost track of the time to such an extent that I decided to remind him in a friendly way that it was now 1880. At that moment, he seemed to come to his senses, and he decided to take his leave.

CAEN. — Les Facultés et la rue Saint-Sauveur.

Fig. 3.3 Poincaré began his university career in Caen (Normandy). The University of Caen was founded in 1432 and was at that time located in the Saint-Sauveur quarter. The buildings were destroyed by Allied bombardment on July 7, 1944, along with most of the old city

Considering the array of exciting ideas that Poincaré would publish in his memoir [Poincaré 1881] and a little bit later on Fuchsian functions, this preoccupation is understandable. In his memoir, Poincaré developed the qualitative theory of ordinary differential equations as we know it today, that is, the classification and nomenclature of the singular points—now usually called critical points—node, focus, saddle, and centre point; the notion of the map of a transversal into itself (Poincaré map); and the concept of limit cycle, which was formulated later as the Poincaré–Bendixson theory. The treatment in [Poincaré 1881] also gives details about the number of possible limit cycles and an analysis of certain difficult cases. The memoir was the start of a completely new approach to the field of ordinary differential equations.

3.3   Automorphic Functions: Contacts with Fuchs and Klein

Poincaré's interest in differential equations and applications led him to develop mathematics along lines closely related to both algebra and complex analysis. An article by Fuchs inspired him to write to the author. Somewhat later, he engaged in an extensive correspondence with Felix Klein. Most of his papers on automorphic functions would appear in these years (1881–1884), but Poincaré retained an interest

Fig. 3.4 Lazarus Immanuel
Fuchs (1833–1902) taught at
Berlin, Greifswald,
Göttingen, and Heidelberg.
He was interested in linear
differential equations with
singularities

in the topic. He published an important paper on uniformization in 1907. Indeed, his
publications on automorphic functions fill nearly a complete volume of his collected
works. Translations of the papers in English, together with an introduction, can be
found in [Poincaré 1985].

In this section, we shall describe Poincaré's contacts with Fuchs and Klein. Some
of the more technical aspects of automorphic functions are described in Chapter 8.

Contacts with Fuchs, 1880–1881

The German mathematician Lazarus Immanuel Fuchs (1833–1902), see Figure 3.4,
studied in Berlin under the supervision of Kummer and Weierstrass. He held
positions in Berlin, Greifswald, Göttingen, and Heidelberg, where he lectured from
1875 to 1884. He returned to Berlin in 1884 to succeed Kummer at the University
of Berlin. The contacts with Poincaré took place when Fuchs was at Heidelberg.
Fuchs was interested in the characteristics of complex-valued solutions of the linear
ordinary differential equation (ODE)

$$\frac{d^2y}{dz^2} + P(z)\frac{dy}{dz} + Q(z)y = 0,$$

with $P(z)$ and $Q(z)$ rational functions of the complex variable z. The functions
$P(z)$ and $Q(z)$ have poles (singularities) at isolated points of the complex plane
$\mathbb{C}$; the strategy is then to look for local series expansions of the solutions in a
neighbourhood of the poles. It is known that the so-called index equation plays a part
in this process. Fuchs begins with two independent solutions $f(z)$ and $g(z)$, defines
the quotient $f(z)/g(z) = \eta(z)$, and asserts in [Fuchs 1880] that the inverse of $\eta(z)$

is a meromorphic function (meromorphic means analytic with the exception of a number of isolated points at which there are poles). The first paper by Fuchs on the subject [Fuchs 1880] presents an incomplete treatment; he excludes, for instance, the presence of logarithmic terms.

Earlier, Fuchs had carried on a correspondence with Hermite on differential equations with singularities. In 1878, the Académie des Sciences offered a prize for the best treatment of the following problem: "To improve in a significant way the theory of linear differential equations of one independent variable." The formulation came from Charles Hermite, who was familiar with the questions raised by the work of Fuchs; the deadline for contributions was the last day of 1880. Hermite, as a prominent member of the jury, had to evaluate the contributions. Such a prize question was one in a long series of scientific competitions in the eighteenth and nineteenth centuries sponsored by various European learned societies and by the nobility. Usually, a prize competition was announced, with a jury, and a deadline for submissions. To ensure anonymity, the contributions were to be submitted in sealed envelopes along with a short sentence or phrase for later identification of the author. The prize competitions focused the attention of a large number of scientists on a major scientific problem. In this way, they played an important role in stimulating research.

In March 1880, Poincaré submitted his article to the Académie, which he then followed up with a revision and a number of supplements. In [Gray and Walters 1997] one can find an extensive description of these contributions. The prize, finally, went to Halphen; Poincaré obtained an honourable mention. Around the same time, on May 29, 1880, Poincaré wrote to Fuchs:

> I have read with great interest the remarkable treatise that you had included in the last issue of the *Journal de Crelle* with the title "Über die Verallgemeinerung des Kehrungsproblems." I hope you will grant me, dear sir, to request from you certain clarifications on this subject. [Here follows a discussion of possible hypotheses.]
>
> However, one could have made a thousand other suppositions. I have to confess, dear sir, that these thoughts have raised with me some doubts about the generality of the results that you have published and I have taken the freedom to approach you about it, hoping that it will not trouble you to clear this up.

In fact, Poincaré demonstrated in this letter that Fuchs had discussed only a special case of his problem. An exchange of letters followed that can be found in [Poincaré 2012] and also in [Poincaré 1916, Vol. 11]. Poincaré wrote in French, and Fuchs replied in German, but that does not seem to have caused any difficulties in communication. In his first letter from Heidelberg, dated June 5, 1880, Fuchs even excused himself for writing in German with the remark that he was certain that Poincaré would have no problem with reading German, the language in which he, Fuchs, was able to express himself most clearly. Fuchs, despite being 21 years Poincaré's senior, consistently maintained a tone of friendship and interest, even when it began to become clear that his young French correspondent was developing an approach that was quite different from his own and more complete. On June 12, 1880, Poincaré wrote to Fuchs from Caen:

I find in the case that there are two singular points only, that the function you have introduced has very remarkable characteristics, and because I intend to publish the results I have obtained, I am asking your permission to call them Fuchsian functions; for it was you who discovered them.

Later, this bestowal of the name "Fuchsian" caused difficulties between Poincaré and Felix Klein. In the meantime, Poincaré analysed many other cases, discovering connections with various special functions, such as elliptic, hypergeometric, and zeta Fuchsian functions.

In 1908, toward the end of his life, Poincaré returned to these discoveries in an essay called "The Invention of Mathematics." He recalled that when he was working in Caen, he had taken part in a geological excursion organized by the École des Mines in Normandy:

The events of the journey made me forget my mathematical work. When we arrived at Coutances, we boarded an omnibus to take us where we would set out for a walk, and just as I put my foot on the step, the idea came to me, though nothing in my former thoughts seemed to have prepared me for it, that the transformations I had used to define Fuchsian functions were identical to those of non-Euclidean geometry. I could not verify it, I had no time to do so, I took up the conversation I was engaged in, but I felt absolute certainty at once. When I got back to Caen I verified at leisure, to satisfy my conscience, the result that I had kept in mind.

The omnibus of Coutances is depicted in Figure 3.5. On March 20, 1881, Poincaré wrote from Caen his last letter to Fuchs with a summary of his results:

I have continued with the functions that I named after you and I hope to publish my results shortly. These functions contain as a special case the elliptic functions and also the modular function. With these and other functions that I called zeta Fuchsian, one can solve:

1. All linear differential equations with rational coefficients that have three singularities only, two finite and one infinite.
2. All second-order equations with rational coefficients.
3. A large number of equations of various orders with rational coefficients.

At this stage, the original article by Fuchs that had inspired Poincaré had disappeared from sight. Poincaré was creating a completely new area of mathematics, that of automorphic functions, which would soon attract the attention of mathematicians interested in the general theory of complex functions: mathematicians from the school of Riemann.

Correspondence with Klein: 1881–1882

The middle of 1881 saw the beginning of an extensive exchange of letters, a total of 26 altogether, with Felix Klein (1849–1925) [Poincaré 2012], [Poincaré 1916, Vol. 11]. In the course of his career, Klein (see Figure 3.6) was attached to a number of universities, beginning with Erlangen. During the time of his correspondence with Poincaré, he was lecturing on geometry at Leipzig (1880–1886); from 1886 he

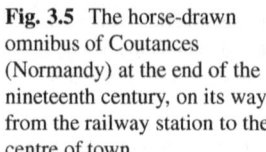

Fig. 3.5 The horse-drawn omnibus of Coutances (Normandy) at the end of the nineteenth century, on its way from the railway station to the centre of town

occupied a chair in Göttingen. Klein was primarily interested not in differential equations themselves but in using such equations to define transformations of complex functions that leave those functions invariant. One of the simplest types is the periodic functions, for which the transformation is the translation of the independent variable by the period. Consider as an example the periodic function $\exp(2\pi i z)$. This function is invariant under the group of translations $z \mapsto z + n$, $n \in \mathbb{Z}$. Discrete and continuous transformations play an important part in the analysis of differential equations carried out by Klein and Poincaré. A prominent role is played by conservative differential equations, which are equations describing physical systems in which the energy is conserved. Such conservative equations usually contain whole families of periodic functions. It is interesting to observe the mixture of mathematical and mechanical thinking in the description of these functions. In this context, Poincaré wrote, in his first supplement to his paper for the prize competition of the Académie in 1880, reprinted in [Gray and Walters 1997, p. 35], referring to the Euclidean and non-Euclidean geometries of Lobachevsky:

> What now is a Geometry? It is the study of a group of operations formed by the displacements of a figure without changing its shape. In Euclidean geometry, this group consists of rotations and translations. In the pseudo-geometry of Lobachevsky, it is much more complicated.

Fig. 3.6 Felix Klein

The first letter of Klein to Poincaré, on June 12, 1881, contained the news that he had read some of Poincaré's papers on Fuchsian functions. Klein went on to describe the sorts of results he had obtained, explaining a number of elementary aspects. On June 15, 1881, Poincaré responded courteously from Caen:

> Your letter shows that you have obtained some results in the theory of Fuchsian functions before I did. I am not at all surprised about that, since I know how well informed you are regarding non-Euclidean geometry, which is the real key to the problem that concerns us. ... When I publish my results, I will do justice to you in this respect.

Poincaré continued by posing a number of questions, and he made a remark on Klein's use of terminology regarding modular functions. Klein answered immediately, on June 19, and it becomes clear what was bothering him:

> I reject the appellation "Fuchsian functions," although I understand quite well that it was through the work of Fuchs that you got these ideas.... I do not deny the great merits of Mr Fuchs in other branches of the theory of differential equations, but exactly in this area his work leaves much to be desired; for the only time he explained modular elliptic functions in a letter to Hermite, he slipped over a fundamental error, which Dedekind later reviewed too leniently.

Klein also listed all the many mathematicians from Riemann's school who were collaborating with him in researching the theory of discrete and continuous transformation groups along with their various contributions. Poincaré's letters to Klein give the impression that he was not aware of the work in the Riemannian school, but he wrote that he would likely be able to reconstruct those ideas. In his reply to Klein on June 22, 1881, he wrote:

> Regarding the name Fuchsian functions, I will not change that. The respect I have for Mr Fuchs prohibits me from doing so. Apart from this, it is true that the point of view of the

mathematician in Heidelberg is completely different from yours and mine. It is also certain that his work served me as the starting point and the basis for everything I have done in this theory.

Klein answered on June 25 rather haughtily:

You would speak differently about F. if you knew the whole literature well.

In his letters, Klein strikes a note that is more that of a tutor than of a colleague, but Poincaré seemed not to mind. He appreciated the dialogue, perhaps because he had no colleague nearby with whom he could discuss such topics. With respect to Fuchs, Poincaré answered on June 27, 1881:

Regarding Mr Fuchs and the name Fuchsian functions, it is clear that I would have chosen another name if I had known the work of Schwarz. But I learned about this only from your letter, after publication of my results, so that I can no longer change the name without failing in consideration for Mr Fuchs.

He added a number of questions and remarks on the paper that Klein sent him. Letters 7, 8, and 9, dated July 2, 5, and 9 (1881), contain extensive mathematical discussions in which the complex analysis of Riemann increasingly plays a part. In the meantime, Poincaré had taken the step of naming a certain function class after Klein. About this, Klein wrote from Leipzig on July 9 (letter 9):

I was somewhat surprised about the name you have attached to this function class, for I did nothing more than to note the existence of this group. As far as I am concerned, I will use neither the "Fuchs" nor the "Klein" indication, but keep to my "functions that contain linear transformations."

As it turned out, in mathematics one today follows Klein in this, and only here and there does one find the term "Fuchsian" used to refer to an automorphic function. However, "Fuchsian groups"—a term also coined by Poincaré—survives as the name of a class of groups that are important in the theory of modular forms. The next letter is from December 1881 and again comes from Klein. He had entered a difficult period of his life. He was suffering from asthma and depression. He had seen in the journal *Comptes Rendus* the fundamental novelty of Poincaré's latest results and how quickly they followed one another. On December 4, 1881, Klein sent a letter containing a proposal:

At the same time that I congratulate you on your far-reaching results, I would like to make a proposal that will satisfy both your and my interests. I would like to ask you to send me a short or longer treatment for the *Mathematische Annalen*, or if you cannot find the time to bring out such a thing, to send me a letter in which you indicate the main lines of your point of view and results. I would then append to this letter a note in which I explain my point of view of the entire field .... Of course I would show you this note for your approval before it goes into print.

Klein could make such a proposal because he was editor-in-chief of the *Mathematische Annalen*.

On October 19, 1881, Henri Poincaré was appointed to the post of Maître de Conférences at the Sorbonne. He replied to Klein on December 8 from Paris, Rue

Gay-Lussac 66, that he would write the paper that Klein had suggested. Klein replied enthusiastically on December 10, proposing that Poincaré's article occupy 16 pages of the *Mathematische Annalen*, and that he submit his article as soon as possible so that everything could be published in March 1882. He did not explain why he was in a hurry. Poincaré might well have assumed that it was just nervous energy.

Already on December 17, 1881, Poincaré submitted the review paper that he had promised, and indeed, it appeared in early 1882 [Poincaré 1882]. Klein thanked him for his contribution on January 13, 1882, and sent Poincaré the note that he would attach to the paper. At the same time, he informed Poincaré that he would produce a short paper of his own containing a few results in progress. The tone of Klein's note, however, was even more forceful than his earlier remarks in his letters. A few quotations:

> The investigations that Mr Schwarz and I published a long time ago in the field under consideration deal with Fuchsian functions, about which Mr Fuchs has not published anything.

And after another series of such remarks, we see a real staking out of territory, incorporating the work of Poincaré:

> Perhaps it is correct on this occasion to add to these small remarks that all research that is discussed here, both what is geometrical in reasoning and the more analytic work that is connected with solutions of linear differential equations, is based on Riemann's ideas. The coherence is even greater because one can state that in the research of Mr Poincaré, what really counts is the further continuation of the general complex function program formulated by Riemann in his dissertation.

How did Poincaré react to this? He wrote a short letter (number 15 in the series) with the remark that he had no desire to change anything in Klein's note, but that he wished to add a few lines to his own article for better justification of the appellation "Fuchsian." This he did a bit later, in the letters dated March 28 and 30, 1882.

The discussion caught Klein in a difficult period, or perhaps there was a connection between his difficulties and the exchange with Poincaré. In looking back at the development of the theory of complex functions [Klein 1924], reproduced in [Poincaré 1916, Vol. 11, pp. 27–28], Klein recounted that he had been ill in the period 1881–1882 and that following medical advice, he had gone in March 1882 to Norderney, a seaside resort on the German part of the North Sea coast. Because of the bad weather, he stayed only eight days, but notwithstanding the short time and his asthma attacks, he formulated an important theorem, which he called "Zentraltheorem." Returning via Düsseldorf, he wrote down the theorem and a proof in a few days, after which he sent it off for publication in the *Mathematische Annalen*. Copies of the manuscript were sent to Schwarz, Hurwitz, and Poincaré. Every researcher knows the feeling of elation at having understood and solved a difficult problem, formulated it well, and sent it off for publication. Klein's proof, however, was not complete. Both Schwarz (see Figure 3.7) and Poincaré expressed doubt and pointed out holes in it; at first, Schwarz did not even believe that the theorem was true. Imagine Klein's annoyance when Poincaré wrote on April 4, 1882:

Fig. 3.7 Hermann Schwarz,
student of Weierstrass

Thank you very much for your last note, which you were so kind to send me. The results that you mention do interest me, and I will tell you why. I found them already some time ago, but without publishing them, because I wanted to clear up the proof somewhat. That is why I would like to know yours when you, from your side, have clarified this.

It was disarmingly honest, typical for Henri Poincaré, but Klein wrote in his recollections, reproduced in [Poincaré 1916, Vol. 11, pp. 27–28]:

How and how long he knew this, he never informed me. It is understandable that this created a certain tension in our relationship.

It is doubtful whether Poincaré realized the importance of this exchange for Klein. The latter was in a disadvantaged position, as anyone would be who had worked on a problem while unbeknownst to him, Poincaré was working on it at the same time. The two had no further discussion about Klein's proof of the theorem.

Back now to the *Mathematische Annalen* and Poincaré's reaction to the note added by Klein. Poincaré wrote on March 28 that he did not want readers of the *Mathematische Annalen* to get the impression that he had wronged someone. The addendum that he sent to Klein (letter 17) on March 30 offered a detailed defence of his decision to name the functions after Fuchs and Klein, while praising the work and the achievements of Schwarz and Riemann.

Klein replied furiously on April 3, 1882, informing Poincaré that he would insert Poincaré's addendum in the *Annalen* with a postscript again reaffirming his own point of view. He concluded by remarking that as far as he was concerned, the altercation over the naming of functions was over, and he expressed the hope that the two of them would maintain regular contact about questions of mutual interest.

Poincaré answered immediately, on April 4. He began ironically:

> You state that on behalf of science you wish to end a sterile debate, and I can only congratulate you on that decision. I know that this will not cost you much, since you are to have the last word, as you know, in your note added to my last piece of writing. Still, I am grateful to you for this.

After a number of remarks on the merits of Fuchs and about questions connected to mathematical problems, he concluded as follows.

> I hope that the quarrel we had because of a name, in this case carried out courteously, will not change our good relationship. In any case, I do not resent it that you initiated the attack; I hope that you will also not take it amiss that I defended myself. By the way, it would be ridiculous to prolong the discussion about a name, "Name ist Schall und Rauch," and when all has been said, I do not care. Do as you wish; from my side I shall do what I wish.

The quotation "Name ist Schall und Rauch" (a name is sound and smoke) is from Goethe's *Faust*, and it is doubtful, to say the least, that Klein was pleased about it. For him, names and reputations, as well as proper acknowledgement of priority of discoveries made by himself and his close colleagues, were of essential importance. There are six subsequent letters containing detailed mathematical discussions, written in a friendly tone. The letter from Poincaré of September 22, 1882, is the last, and thereafter, the correspondence between the two researchers ceased, although both of them would be working on automorphic functions for many years to come. In 1906, for instance, Poincaré submitted a paper on the uniformization of analytic functions to Mittag-Leffler, the editor of the journal *Acta Mathematica* [Poincaré 1999, letter 232].

Was there too much emotion in the quarrel? Mittag-Leffler, who often travelled through Europe and acted as a trusted representative of many mathematicians, wrote to Poincaré [Poincaré 1999, letter 17] that he had visited Schwarz, who "was beside himself with fury" about Poincaré's choice of names. Fortunately, relations between Poincaré and Schwarz improved when Schwarz visited Paris in April 1884 [Poincaré 1999, letter 29]. One may advocate the rational attitude that quarrels can be settled by each side saying, "Let us agree to disagree," but that denies emotions and ambitions that cannot be gainsaid. At this time, Schwarz aspired to a professorship at the University of Berlin. For Klein, the choice of names was clearly very important, but there was probably yet another element in play. The correspondence that we have been describing was between a well-established professor in Leipzig and a neophyte young mathematician from Caen. The latter turned out to be a mathematical prodigy, a genius who was always far ahead of everybody else working on the same topics. While Klein and his students were working on special problems, Poincaré formulated the theory from the outset in great generality; see also the comments in [Freudenthal 1954].

Felix Klein was an eminent mathematician for whom fame and prestige counted heavily. His luck was against him when he was working on topics in which Poincaré was also interested. The confrontation with Poincaré must have cost him emotionally, but it is remarkable, given all the dissonances over the naming of functions, how courteous and civilized the correspondence between the two great mathematicians actually was.

Chapter 4
Career in Paris

The marriage of Henri Poincaré and Louise Poulain d'Andecy took place on April 20, 1881, in Paris. Their first home in Paris was at Rue Gay-Lussac 66. Later, they moved to Rue Claude Bernard 63. In 1887, their first child was born, a daughter, Jeanne; two daughters were born a few years later, Yvonne in 1889 and Henriette in 1891. The year 1893 saw the birth of their fourth and last child, a son, Léon.

There are only five letters in the Poincaré archive [Poincaré 2012] with dates in 1886 and 1887 from Henri to Louise (see Figure 4.1), and none from Louise to Henri. The letters are very affectionate, opening, for instance, with, "My dearest darling, I have received your letter. I adore you." And ending with, "My dearest darling, I love you very much, very much" [Poincaré 2012, Louise 0ab]. The letters contain accounts of meetings with various people, the quality of hotel rooms, and interesting gossip. Poincaré liked to travel, and Louise accompanied him now and then. She had been raised in a well-known intellectual family, and knowing the special requirements of such a milieu, she provided Henri and their children a happy and safe home. Together, they visited exhibitions and concerts. Poincaré was especially fond of symphonic music.

In the Poincaré family, as in all families, there was birth and there was death. In the year between the births of Henri and Louise's third and fourth children, Henri's father, Léon Poincaré, died, on May 21, 1892. Thus he lived to see his three granddaughters, but not his grandson.

Henri's mother, Eugénie, died five years later, on July 15, 1897. Henri was shattered by this event, and for several months thereafter, he did not answer any letters and kept himself confined to his family. In a note to the Swedish mathematician Gösta Mittag-Leffler [Poincaré 1999, letter 143], he wrote on July 31 of that year that he was unable to work and that he could not discuss any requests for refereeing or editing papers. On August 3, Mittag-Leffler replied with understanding and sympathetic condolences. It took until October 11 for Poincaré to resume his correspondence.

In August 1897, the First International Congress of Mathematicians took place in Zurich. Poincaré was a member of the organizational committee and had been asked to present the opening address. In fact, he had already submitted the text

F. Verhulst, *Henri Poincaré: Impatient Genius*, DOI 10.1007/978-1-4614-2407-9_4,
© Springer Science+Business Media New York 2012

Fig. 4.1 Henri Poincaré and Louise Poulain d'Andecy at the time of their marriage in 1881

of his address before his mother died. Poincaré felt unable to attend, and a Swiss colleague, Jérôme Franel, read his lecture on the relationship between pure analysis and mathematical physics, "Sur les rapports de l'analyse pure et de la physique mathématique" (on the reciprocal relations existing between pure analysis and mathematical physics). There were around 35 plenary lectures, one of them by Felix Klein. The text of Poincaré's opening address is reproduced in the philosophical book [Poincaré 1905b]; it is discussed in this book in Section 6.5.

Poincaré was appointed to the position of *maître de conférences* in the Faculté des Sciences at the Sorbonne in Paris on October 19, 1881. This type of position had only recently been instituted; the idea was to attach to the holder of a regular chair a sort of tutor or coach, a qualified lecturer who looked after his students, gave them problems and exercises, and corrected their papers. Apart from those tasks, this *maître* could follow his own independent line of research.

Until his death in 1912, 31 years later, Poincaré continued to live and work in Paris, exhibiting extraordinary creativity and productivity in many fields. In this chapter we will give a bird's-eye view of his activities. A number of aspects will receive special attention in separate sections and chapters, while more technical aspects will be treated in the second part of this book.

4.1   Sketch of a Scientific Career

The position of *maître de conférences* at the Sorbonne was followed very quickly by other appointments. On November 6, 1883, Poincaré obtained the additional position of tutor at his former place of education, the École Polytechnique; he would keep this position until March 1, 1897.

Although this tutorship was useful for making professional contacts, the appointment of March 16, 1885, as professor in the chair of physical and experimental mechanics at the Sorbonne provided many more possibilities. Poincaré was then 30 years old. However, the experimental side of the assignment suited him much less than the theoretical side, but conveniently, another chair soon became vacant. In the summer of 1886, he succeeded Gabriel Lippmann (1845–1921) in the chair of mathematical physics and probability; Lippmann moved to experimental physics and was awarded the 1908 Nobel Prize in physics for his development of photographic processes.

Regarding Poincaré's lectures and his attitude toward experiments, there is a revealing account from Maurice d'Ocagne (1862–1938). D'Ocagne was a gifted mathematician and civil engineer who entered the École Polytechnique in 1880. From 1893 he was also employed as a tutor at the school; he wrote literary essays under the pseudonym Pierre Delix as well as a comedy that became very successful in Paris. He writes about Poincaré's lectures on physical and experimental mechanics (cited in [Bellivier 1956]):

> One cannot say that Poincaré was a great lecturer; he dominated his audience from on high and was not gifted as an orator. The demonstrations left also something to be desired; one had clearly a feeling that he had never given the least attention to the construction and use of the instruments. When one of these instruments was placed on his table during the lecture, there was nothing more amusing than to follow the timid way in which he tried to employ it. He approached it several times with an astonished expression, as if this material realization had failed to adapt itself to the well-understood purely schematic picture that he had in mind. He tried to grasp it by some screws that he accidentally touched, and ... he then gave up.

Perhaps the paralysis that Poincaré had suffered when he was five years old made him uncertain and impeded somewhat his physical coordination.

The appointment in mathematical physics, which was more theoretical, gave him the possibility to expand in many fields, both as a lecturer and in research. Each year, he tackled another subject for his lectures, usually in connection with his own interest at the time. These lectures are summarized in books based on notes written down and edited by his students, the actual publishing being organized by the *Association amicale des élèves et des anciens élèves de la Faculté des Sciences*. Some of these students became famous later on. The final text in each case was checked and approved by Poincaré.

In Section 4.5 we present a survey of these topics from mathematical physics, and in Chapter 11 we discuss some of them in more detail. The lectures are concerned primarily with theoretical physics, but with a solid mathematical foundation. Mathematical physicists restrict themselves usually to either the mathematical background of physical phenomena or to the physical consequences of the mathematical theory. Poincaré was also in this respect exceptional, for he developed fundamental mathematics while simultaneously considering the physical implications. As we have seen, his gifts were not only mathematical, and already as a child he exhibited a wide range of interests.

At an early stage in his career, Henri Poincaré's creativity must have been widely recognized, for on January 31, 1887, at the age of 32, he became a member of the Academy of Sciences. He received 34 votes, to 24 for Colonel Amédée Mannheim of the École Polytechnique, his most serious competitor. Mannheim was much older and was at the time not very fond of the student Henri Poincaré.

For many scientists, an appointment to the Academy of Sciences is the crowning conclusion of a successful scientific career, but for Poincaré, it was just the beginning. King Oscar II of Sweden had offered a prize for the best scientific essay on a choice of several topics, one of which was the stability of the solar system. For his contribution, Poincaré was awarded the first prize in 1889; see Chapter 5. His friend Appell came in second.

The renowned astronomer Félix Tisserand (1845–1896) died in 1896, and Poincaré became his successor in the chair of mathematical astronomy and celestial mechanics. In the period 1892–1899, three volumes by Poincaré appeared on new methods in celestial mechanics [Poincaré 1892]; see also Section 9.3 of this book. Although these books have as a starting point celestial mechanics, the study of the motion of point masses in a gravitational field, they can be considered the first modern treatment of dynamical systems. The concepts and methods developed in these books have a formulation and generality that go far beyond the specific problems in celestial mechanics used to illustrate the theory. Fifty years earlier, Carl Gustav Jacobi had made an enormous step forward in the field of dynamics. Making use of Jacobi's work, Poincaré gave a completely new approach involving variational and linearization methods, integral invariants, and periodic and other types of special solutions.

4.2   Contacts and Travels

For the modern scientist, travel to scientific conferences and research centres, along with regular, often daily, contact with colleagues for stimulating discussions and to keep in touch with the latest progress in their field, is a way of life. Around 1900, travel was neither as rapid nor as simple as it is today, and the majority of scientific contact was by exchange of letters. Nonetheless, that era saw the beginning of international conferences, with Poincaré as a regular participant. Moreover, Poincaré was invited to a number of foreign institutions to receive scientific honours and prizes. His memberships in foreign academies and his honorary doctoral degrees are too many to list here. A list of degrees, honours, and decorations can be found in [Lebon 1912].

In 1906, Poincaré was elected president of the Académie des Sciences, to which, as mentioned above, he had been elected in 1887. On March 5, 1908, he became a member of the prestigious Académie française. This society was founded in 1635 by Cardinal Richelieu with its primary mission the regulation and advancement of the French language. Such an institution is almost unique in the world, although academies inspired by some of the ideas of the French Academy have been founded in Spain, Brazil, Russia, and Sweden. There are only 40 members, the "immortals." Poincaré was appointed to chair 24, which had become vacant on the death of the poet Sully Prudhomme. In his welcoming speech, the director, Frédéric Masson, mentioned that Poincaré was already a member of 35 academies. While describing the scientific and philosophical achievements of the new immortal, he emphasized that the Académie française always reserved a place for scientists to facilitate their active collaboration in clarifying the meaning and use of scientific terms and phrases (for an extensive account of Masson's speech see [Lebon 1912]).

We now discuss briefly some of Poincaré's travels and conferences that he attended. We mentioned earlier the first International Congress of Mathematicians, which took place in 1897 in Zurich, where Poincaré's plenary lecture was read by a colleague. During August 6–12, 1900, the second International Congress of Mathematicians took place in Paris, with Poincaré as chairman. There were around 40 lectures, including papers by Hilbert and Mittag-Leffler. The latter discussed the correspondence and contacts between Weierstrass and Sonya Kovalevskaya in his lecture "Une page de la vie de Weierstrass." Hilbert presented his famous list of mathematical problems to be solved in the twentieth century. The 16th problem was "to determine the maximum number of limit cycles of Mr Poincaré's equation $dy/dx = f(x, y)$ with $f(x, y)$ polynomial." Poincaré would mention the problem again in 1908. The international mathematical conferences before the First World War had no common language: the Italians, the French, and the Germans used their own languages; Mittag-Leffler spoke French in France, German in Germany. Poincaré's plenary lecture was called "De l'intuition et de la logique en mathématiques." Except for a few small changes, it is completely contained in the philosophical book [Poincaré 1905b, Chapter 1]. Poincaré emphasized intuition as complementary to analysis and deduction. See Chapter 6 for further discussion.

Two years earlier, in 1900, Poincaré travelled to London to receive the gold medal of the Royal Astronomical Society. On February 9, at a special session of the society, he was addressed by its president, George Darwin. In his eulogy, Darwin pointed out that Poincaré had successfully studied many topics, but his address focused on the theory of tidal motion and on the stability of rotating fluid masses [Lebon 1912], which were subjects on which Darwin himself had worked.

An international congress of arts and sciences was planned in conjunction with the 1904 Louisiana Purchase Exposition, a major international World's Fair held in the United States, in St. Louis, Missouri, to commemorate (one year late) the 1803 purchase of the territory of Louisiana by the United States. In the nineteenth century, there were still very few internationally known artists and scientists in the New World. The organizers of the fair hoped to put on the cultural and scientific world map a country known in Europe mainly as a land of pioneers. In 1903, the Canadian-American mathematician and astronomer Simon Newcomb (1835–1909) was delegated to invite noted Europeans to attend the conference. One of those he invited from France was Henri Poincaré, who gave in St. Louis a remarkable lecture on the principles of mathematical physics, anticipating a number of concepts of relativity. On this occasion he also met the American astronomer and mathematician George Hill (1838–1914), whose work on the differential equations of celestial

Fig. 4.2 Gaston Darboux, one of Henri Poincaré's dissertation advisers and later colleague

mechanics he knew and appreciated. There is an apocryphal story that the university professors who invited him asked Poincaré whom he would like to meet. Poincaré mentioned only one name, Hill, but to their embarrassment, the professors did not know of Hill, who had worked for a long time at a nonacademic institution, the Nautical Almanac Office, and was rather reclusive.

The fourth International Congress of Mathematicians took place in Rome, April 6–11, 1908, with more than a hundred lectures. A novelty of the congress was that for the first time, there were sections devoted to pedagogy and to applied mathematics, including lectures by the British applied mathematician Horace Lamb and the German mathematical physicist Arnold Sommerfeld. Poincaré was not present at the third congress, in Heidelberg, but he went to the Rome congress to present a plenary lecture. Unfortunately, he became seriously ill because of a prostate condition. It was the first sign of the illness that would lead to his death in 1912. He had prepared a lecture on the future of mathematics that was read for him by Gaston Darboux, see Figure 4.2. It is substantially reproduced in [Poincaré 1908a, Chapter 2]. The differences with the original text are discussed in the following subsection. Louise Poincaré travelled to Rome to be with her husband; when he was sufficiently recovered, they returned in stages to Paris, travelling relatively short distances each day.

In 1905 and again in 1910, Poincaré travelled to Budapest. The 1910 journey was on the occasion of David Hilbert receiving the Bolyai Prize from the Hungarian Academy of Sciences. Poincaré was given the honour of reporting on Hilbert's accomplishments. In 1905, Poincaré himself had received the Bolyai Prize (voted on in 1901), with Gustave Rados (1862–1941), member of the academy, as reporter on Poincaré's accomplishments. This was the first time the prize was awarded. Rados (see [Lebon 1912]) emphasized Poincaré's results on automorphic functions while briefly discussing a large number of very different results, ranging from celestial mechanics to topology.

In 1909, Poincaré was invited by the Wolfskehl Committee to give six lectures in Göttingen. This occasion presented an opportunity for him to meet his colleagues Felix Klein and David Hilbert, but he was still suffering from prostate problems. While preparing his lectures for Göttingen, to be given in the week April 22–29, 1909, he wrote to Hilbert [Poincaré 1999, p. 349]:

> There is a point now to which I want to draw your attention. I am still experiencing the consequences of the accident that struck me last year in Rome and I am strongly compelled to take certain precautions. I cannot drink wine, also not beer, but only water. I cannot be present at a banquet, also not at a prolonged dinner.

The trip proceeded as planned. The first three lectures dealt with integral equations and their physical applications; the fourth was on abelian integrals and Fuchsian functions. The fifth lecture discussed transfinite numbers (cardinal numbers and the continuum hypothesis), while the last dealt with the new mechanics. In the first five lectures, Poincaré spoke in German, while in the last, he spoke in French. The new mechanics is what we today call special relativity, and we shall see how this was viewed in 1909 in Sections 4.7 and 11.4.

The year 1911 saw the beginning of a historical series of conferences in Brussels, sponsored by the Belgian industrialist Ernest Solvay. Hendrik A. Lorentz would chair the first five conferences, with, as it was said, "tact and acumen." In 1911 the topic was the theory of radiation and quanta. It was the first and last occasion on which Poincaré and Einstein met. The participants noticed how actively Poincaré participated in the discussions and how interested he was in the emerging quantum theory.

The Rome Lecture on the Future of Mathematics

Poincaré's original lecture of 1908, which can be found on the website of the International Mathematical Union ("Historic IMU/ICM"), was polished and simplified for a wider public in the philosophical book *Science et Méthode* [Poincaré 1908a]; the section on arithmetic was reduced by half. In his talk, Poincaré stressed the use of linear and nonlinear transformations. He expected much from the use of discontinuous groups and Minkowski's geometry of numbers. We discuss here briefly the topics that were omitted in [Poincaré 1908a].

Regarding *differential equations*, there is little knowledge about nonlinear equations. Considering the neighbourhoods of singular points would give a first classification, and using transformation groups may play a useful part, as it does with birational transformations for algebraic curves. Considering first-order equations $dy/dx = f(x, y)$, we do not even know enough about the integrals $F(x, y) =$ constant. We do not know what parameterization of the integrals will also satisfy the differential equation. It would be important for qualitative insight to know the number of possible limit cycles in first-order equations.

Most *partial differential equations* in mathematical physics are linear. Fredholm's theory of infinite determinants has been very helpful here. Together with the variational approach of Dirichlet as advanced by Hilbert, this research will continue. Poincaré states that it will not be easy to combine the two methods, and also he questions whether much can be gained from such a combination.

The theory of *abelian functions* can now be considered complete; some are associated with integrals and algebraic curves, some with integrals and manifolds.

In the *theory of (complex) functions* there is an essential difference between functions of one variable and functions of more than one variable. It is not clear what our knowledge of rational functions of two variables tells us about transcendental functions of more variables. Can the same uniformization as for functions of one complex variable (see Chapter 8) be carried out for more variables?

Poincaré considers *group theory* a very extensive field and restricts himself to Lie groups and Galois groups. Lie group theory has advanced considerably and should now be provided with simpler proofs and classification of results. In Galois theory there has not been much progress. The parallelism with Lie theory should help.

4.3   Paul Appell

This section is devoted to a more detailed discussion of Henri Poincaré's friend Paul Émile Appell (September 27, 1855–October 24, 1930), whom we have met in previous chapters. Paul Appell (see Figure 4.3) was born in Strasbourg, Alsace, a long-disputed territory between France and Germany. His father, Jean-Pierre Appell, was a dyer in Ritterhus, with most of the family working in that business. In 1870, Alsace was annexed by Prussia, and the Appell family moved to Nancy to remain French. Paul attended the Nancy lycée, where, as described earlier, he met Henri Poincaré, with whom he shared a lifelong friendship.

In 1873, Paul and Henri moved to Paris to continue their studies, Paul at the École Normale Supérieure, graduating in 1876 with a first place. In 1881, he married Amélie Bertrand, a niece of the mathematicians Joseph Bertrand and

Fig. 4.3 Paul Appell

Charles Hermite and a cousin of Émile Picard. In 1885 he became professor of mechanics at the Sorbonne in Paris, and in 1892 he was elected to the Académie des Sciences. From 1903 to 1920 he was dean of the Faculté des Sciences of the Sorbonne, and from 1920 to 1925 he was rector of that university. In addition, he held many important public posts during his life. In 1925, he published a biography of Henri Poincaré [Appell 1925a]. In Chapter 7, we cite some of his remarks from that biography. In their later years in Paris, Appell and Poincaré had many occasions to talk together at the Sorbonne, often while walking home after work. Their cooperation did not take the form of joint papers or books, but continuous discussion and exchange of ideas.

Paul Appell wrote hundreds of articles and books on analysis, geometry, and mechanics. His grasp of problems and his approach to their solution were brilliant, but his talent was directed more at solving problems and not so much at developing general theory. This lack of taste for generalization and abstraction is probably the reason why he is not so well known today as some of his contemporaries. His survey of publications by himself [Appell 1925b] is far from complete. We describe some of his remarkable achievements.

In 1880, he constructed the so-called Appell polynomials [Appell 1880]. These are complex polynomials of a very general form. As special cases they contain the Bernoulli, Hermite, and Laguerre polynomials. The expansion of analytic functions implicitly defined by functional and differential equations can be performed in terms of Appell polynomials. Modern extensions, already explored by Paul Appell, are two-variable polynomial expansions that are analogues of classical (Jacobi) orthogonal expansions.

An ingenious approach combining geometry and mechanics involves a homographic transformation of the plane into itself. This Appell transformation can be used [Appell 1891] to determine the forces working on a material point in the plane that produce conic sections as trajectories (Bertrand's problem).

The Gibbs–Appell equations, considered earlier by Gibbs, form an alternative way to describe both holonomic and nonholonomic systems in dynamics [Appell 1900]. For the holonomic case, they are equivalent to the Lagrange equations of mechanics, but since they include the nonholonomic case (systems with constraints, for instance a top moving on a surface), they represent a very general formulation of the differential equations of classical mechanics (see also [Appell 1921, Vol. 2]).

A very complete treatment of mechanics and related mathematics, perhaps the most complete in the scientific literature of the twentieth century, can be found in Appell's four-volume *Traité de Mécanique Rationelle* [Appell 1921]. It is also a rich source of references. The first volume deals with the basic theory of statics and dynamics, including many classical examples. The second discusses the theory of systems of mass points, with topics including integral invariants and nonholonomic systems. Apart from the general theory, there are numerous applications to pendulums, collisions, and engineering problems. Volume three considers equilibria and dynamics of continuous media, for instance wave propagation with discontinuities, fluid mechanics, and elasticity. Volume four is of a different nature, focusing on the particular problem of equilibrium states of rotating homogeneous fluid masses with Newtonian attraction of the particles. These problems are relevant to the theory of planet and star formation. Newton had already observed that a fluid mass rotating with constant angular velocity around a fixed axis should produce an ellipsoidal figure with flattening at the poles. Maclaurin showed in 1742 that these ellipsoids are in fact equilibrium figures by applying the laws of hydrostatic pressure; this was followed by many contributions from other scientists. In 1834, Jacobi added the three-axial equilibrium ellipsoid to the possible equilibria. These ellipsoids rotate about their smallest axis. For the stability of the equilibria, the rotation speed is an essential parameter. The volume contains a description of figures found by Poincaré (see also Section 11.2), such as the pear-shaped figure and the halter, together with stability calculations.

4.4   Contacts with Mittag-Leffler

A special relationship developed between Poincaré and the Swedish mathematician Gösta Mittag-Leffler, see Figure 4.4, who studied with Hermite in Paris and with Weierstrass in Berlin. Mittag-Leffler's formative years led him to adopt a rigorous approach to the formulation of mathematical results, so it is no surprise that he did not like Poincaré's style, which was intuitive and informal. But he could not ignore the many brilliant ideas in Poincaré's papers. The correspondence between the two mathematicians began in 1881 and ended only in 1911; see [Nabonnand 1999], and for the complete correspondence, [Poincaré 2012] and [Poincaré 1999], which records 259 letters. For both mathematicians, such a large number of letters

Fig. 4.4 Gösta
Mittag-Leffler

was unexceptional, but the long run of the correspondence, around thirty years, was unusual. The first letters of 1881—Poincaré was still in Caen—are rather formal, but quite soon both of them struck a cordial note; however, the letters contain few personal and intimate remarks, with the exception of letter 143 in [Poincaré 1999], in which Poincaré wrote briefly about his distress on the death of his mother.

At the time of their first contact, Mittag-Leffler was already a well-known professor, one of the best students of Weierstrass. Like Klein, he adopted in his first exchanges with Poincaré a tone that was slightly condescending, but this changed when he learned about Poincaré's achievements. Yet still in 1883, Mittag-Leffler wrote to Weierstrass:

> What do you make of Poincaré's second paper "Sur les fonctions fuchsiennes"? It is indeed regrettable that he is not a graduate of a German University. As full of new ideas as his papers are, they leave, it seems to me, far too much to be desired in their formal presentation.

He made similar complaints in letters to Hermite (see [Nabonnand 1999]), but in retrospect, one can say that trying to force Poincaré to adopt the style of a Weierstrass would have been like harnessing a racehorse to an oxcart. Considering Poincaré's complaints about the teaching of mathematics at the École Polytechnique, such an attempt might even have alienated him completely from mathematics. Notwithstanding the differences in style between the two mathematicians, a long professional and friendly relationship developed.

In May 1882, Mittag-Leffler married Signe Lindfors, a wealthy young lady from Helsinki. The first meeting of the two mathematicians took place on the occasion of the honeymoon of the Swedish pair, which took the form of visiting a number of mathematical institutes in Europe, eliciting from Weierstrass the following observation [Poincaré 1999, p. 68]:

> Mittag-Leffler and his wife were here last week, from Wednesday until Sunday evening. I have met them often. One liked the young lady very much; one also admired her simple but very elegant dresses. Mittag-Leffler made in practice a mathematical journey— Strasbourg, Heidelberg, Göttingen, Leipzig, Halle, Berlin—with the exception of Paris. Certainly interesting for him—I would not say that this was also the case for the young woman.

The couple visited Paris, but—and this was the exception—not only for mathematical reasons. Paris was likely more culturally rewarding for Signe than the other cities the couple visited, and it was also personally gratifying, for here she and her husband first met Henri and Louise Poincaré. Signe and Louise liked each other, as becomes clear from the letters in [Poincaré 1999]. Poincaré travelled several times to Sweden, where he visited Mittag-Leffler, for instance in June 1905.

In 1882, Mittag-Leffler wrote to Hermite, Appell, and Poincaré about a new project. He proposed founding a journal, to be called *Acta Mathematica*, with himself as editor-in-chief and a strong presence of German and French mathematicians among the editors. The king of Sweden, Oscar II, supported the project financially, as did several Scandinavian governments. This turned out to be an important project for mathematics. At that time, the centre for mathematics research was in Germany, where papers were written in German. France also could boast eminent mathematicians with important contributions written in French. However, few Germans read French, and very few Frenchmen read German. Moreover, after the end of the Franco-Prussian War of 1870–1871, during which France had been invaded by the Prussian army, the French became distinctly Germanophobic. Typical was that the avalanche of results in automorphic functions obtained by Poincaré in advance of Klein was not considered a victory for mathematics or science but a victory for France. In the same way, the award of a prize in an international competition to a French scientist was considered a French conquest, a contribution to the glory of France. When Poincaré and Appell won such prizes in 1889, both of them were awarded the Légion d'Honneur.

The result was that the work of mathematicians such as Weierstrass, Cantor, and Schwarz was relatively unknown in France, while the work of Hermite, Darboux, Laguerre, and other French mathematicians was little known in Germany. A new journal from a neutral place like Stockholm could bridge this gap by producing editorial cooperation among mathematicians of various nationalities.

For the enterprise, one can think in retrospect of nobody better equipped for this task than Mittag-Leffler. He had studied in Berlin with Weierstrass and in Paris with Hermite. He had proved himself a very good mathematician, but the essential point to complete this intellectual equipment was that he was a natural diplomat. In Stockholm, he operated skilfully by recruiting the right people for the newly founded university. He kept the Swedish king interested in science, and all over Europe he attracted the best mathematicians as contributors to the *Acta*. The languages most used in the *Acta Mathematica* were French and German. In 1883, Hermite wrote to the Swedish ambassador to France that the *Acta Mathematica* was the first foreign scientific journal to which the French scientific faculties had taken a subscription [Poincaré 1999, p. 19]. No doubt, this letter was suggested by Mittag-Leffler to ensure continuing support from his king.

Another example (of many) of Mittag-Leffler's diplomatic qualities is his efforts to have Poincaré awarded the Nobel Prize in physics. He tried to enlist the support of the Swedish physicists, but these were mainly experimentally oriented, and for instance Arrhenius (1859–1927), both physicist and chemist, found Poincaré too much of a theoretician to support him. As a first step toward creating an atmosphere in which theoreticians would also be eligible for the prize, Mittag-Leffler proposed Hendrik A. Lorentz (1853–1928) as a candidate. Lorentz was undoubtedly the greatest theoretical physicist of his time, and he was awarded the Nobel Prize in 1902, but together with the experimental physicist Pieter Zeeman. This was a turning point for the Nobel committee, but for a long time, they still remained suspicious of theoretically oriented physicists. Mittag-Leffler was not afraid to express strong opinions about topics in which he was definitely no expert. In 1908, he swung the vote in the Nobel committee to give the physics prize to Gabriel Lippmann. The nomination of candidates was prepared by subcommittees, and in a letter to Painlevé [Poincaré 1999, p. 349], he writes:

> It is I who together with Phragmén have given the prize to Lippmann. Arrhenius wanted to give it to Planck in Berlin, but his report, which he was able to carry through the

subcommittee with unanimous support, was so silly that I was able to destroy it. In the end, he got only 13 votes (clearly including Retzius), whereas I got 46 votes. Two members of the subcommittee declared that after they heard me, they changed their opinion and voted for Lippmann. I would have had nothing against sharing the prize between Wien and Planck, but to give it to Plank alone would be rewarding ideas that are still obscure and that have to be checked by mathematics and experience.

Gustaf Retzius (1842–1919) was a well-known member of the Swedish Academy of Sciences and a specialist in medicine; Wilhelm Wien (1864–1928, Nobel Prize 1911) and Max Planck (1858–1947, Nobel Prize 1918) were prominent physicists. If Poincaré had not died prematurely in 1912, Mittag-Leffler would have continued his efforts.

Part of the success of the *Acta* was also due to Poincaré, who contributed many long memoirs on automorphic functions, celestial mechanics, and other topics. Other French mathematicians made contributions, including Paul Appell and Gaston Darboux. There appeared in the *Acta* translations from German into French of papers by Weierstrass and Cantor. Altogether, Gösta Mittag-Leffler's contribution to the organization of science in Sweden, and in Europe in general, was considerable and permanent. The correspondence reproduced in [Poincaré 1999] also reflects the private and sometimes painful discussions about the awarding of the prize by King Oscar II on the occasion of his birthday; see Chapter 5. After the prize episode, the letters in [Poincaré 1999] are more and more concerned with the daily problems of editing manuscripts for the *Acta* and the discussion of nominations for prizes and positions.

4.5   Lecture Notes and Students

The notes of Poincaré's lectures [Poincaré 1890a] contain a remarkably wide range of topics in mathematics and mathematical physics. The three volumes of *Les Méthodes Nouvelles de la Mécanique Céleste* [Poincaré 1892] are not included in this list, for they are not regular textbooks. These volumes contain a number of applications to celestial mechanics, but their main content is the development of a very general and fundamental theory of dynamical systems; see Section 9.3. On the other hand, the three volumes of the *Leçons de Mécanique Céleste* were written for practical use by astronomers; they are lecture notes. In the second volume of the *Leçons*, the perturbation function is developed with the method discussed in Chapter 6 of the first volume of [Poincaré 1892], but now for actual solar system models. This involves subtle complex analysis. Poincaré notes that in the *Leçons*, the mathematics and rigour of [Poincaré 1892] are absent but that he aims at calculations with high precision, in fact higher than is realistic with regard to the observations of that time but expecting that observations will improve, requiring more advanced calculations. He also notes that together with his own books on mathematical methods [Poincaré 1892], the older books of Tisserand (1845–1896) contain an excellent introduction to celestial mechanics; Poincaré refers to this and warns that he does not intend to duplicate the chapters of the four volumes of [Tisserand 1889].

The text on the theory of cosmogony that appeared in 1911 was also a special topic for Poincaré; it contains an extensive critical analysis of all cosmogonic hypotheses up to the year of publication. He strongly advocates the nebula hypothesis of Laplace. More details can be found in Chapter 11 of this book. Darboux [Darboux 1913] notes that this analysis stops just before a new branch of astronomy, astrophysics and in particular spectroscopic analysis, emerges, but that it contains a very precise evaluation and summing up of all the classical theories of cosmogony. In the preface, Poincaré writes that we do not know enough theory and have not made enough observations to have a serious hope of developing an acceptable theory. But "if we were so reasonable, if we would be curious without impatience, we would probably never have created science, and we would always have been happy with living our petty life." An interesting appreciation of Poincaré's work and ideas by Ernest Le Bon was added to the second edition of the cosmogony. Some of the other topics, such as the propagation of heat and the figures of equilibrium of rotating fluid masses describing bifurcation phenomena, will be discussed in Chapter 11.

Apart from the fundamental celestial mechanics books [Poincaré 1892], the bibliographic details of the lecture notes are given in [Poincaré 1890a]. Here we list the topics of the lecture notes:

1. Celestial mechanics (the *Leçons*)
2. The mathematical theory of light, two volumes
3. Electricity and optics, two volumes (Maxwell theory and Hertz oscillations, wireless telegraphy)
4. Thermodynamics
5. The theory of elasticity
6. Vortical motion ("tourbillons")
7. Electrical oscillations
8. Capillarity
9. Analytical theory of the propagation of heat
10. Probability
11. Potential theory
12. Kinematics and fluid mechanics
13. Equilibrium figures of a fluid mass
14. Cosmogonic hypotheses

The idea of the lectures at the Sorbonne and the corresponding books was to present mathematical physics in the spirit of Laplace and Cauchy. This means that one starts with a few physical hypotheses to develop the corresponding theory as completely as possible, i.e., to obtain a theoretical explanation of the physical phenomena and to compare it with experiment. The treatment of celestial mechanics and potential theory is typical for such an approach, since the physical nature of the phenomenon, gravitation, is well described by classical mechanics. For newer parts of physics, where the foundations are still a subject of investigation and discussion, this is more difficult, but it still holds as an ideal. In this respect, Poincaré is quite critical of Maxwell's presentation, which he says is admirable for its ideas but at the same time is unfocused and lacking in a systematic construction of the theory. Darboux [Darboux 1913] agreed with this point of view, but noted that Joseph Bertrand (1822–1900) did not. Darboux quotes Poincaré as saying about Maxwell's work:

> Perhaps a day will come when the physicists will not be interested in the questions that are not accessible to positive methods and will leave them to the metaphysicians. This day has not come; humanity will not so easily resign itself to remain in ignorance about the foundations of the field of inquiry.

In Section 4.6, we present Poincaré's criticism of Maxwell in more detail.

Because of Poincaré's interest in celestial mechanics, one might obtain the superficial impression that his work on differential equations was mainly concerned with ordinary differential equations. However, as the titles of the lecture notes already suggest, in a considerable number of the lectures Poincaré used and developed partial differential equations. There are many original results there, for instance on the Laplace and Poisson equations, balayage methods, and wave equations; see Chapter 11.

Most remarkable is the enormous quantity of work produced. The actual periods of lecturing cover very different topics, for example:

1. 1888–1889: electricity and optics, first part
2. 1889–1890: electricity and optics, second part
3. 1891–1892: second semester, vortex motion ("tourbillons")
4. 1893–1994: first semester, theory of heat
5. 1893–1994: second semester, probability

The influence of these lectures on the education of French students at that time must have been enormous. The notes were edited by students (see [Poincaré 1890a]) and finally approved by Poincaré. Some of the students made names for themselves later; we mention Émile Borel (1871–1956), Jules Drach (1871–1949), and René Baire (1874–1932). Interestingly, Tobias Dantzig (1884–1956) was studying mathematics in Paris at that time. He was born in Latvia, studied and married in Paris, and emigrated in 1910 to the United States. He received his doctorate in mathematics in 1917 and was the father of the American mathematician George Dantzig (1914–2005), who became famous for his work on linear programming.

Around 1900 and even until the middle of the twentieth century, writing a doctoral dissertation in France was a solitary business, with the doctoral student having only occasional contact with a supervising professor. When the thesis was completed (or if the student thought such was the case), it was submitted to the analysis and judgment of a committee. If the thesis was given the grade "très honorable," there was a good chance for its author of a subsequent career at the Sorbonne. There are three mathematicians for whom Poincaré clearly was important as a supervisor of their doctoral work. Well known is Louis Bachelier (1873–1946), who in 1900 presented a thesis on the theory of speculation (see Figure 4.5). With

Fig. 4.5 Louis Bachelier
(1873–1946), one of the
founders of financial
mathematics, student of
Poincaré

its results on the probabilistic consequences of buying and selling equities, it was far ahead of its time. In the field of financial economics, Bachelier's work was recognized only after his death. In his report for the committee, Henri Poincaré called the topic unusual but noted that the work was of high quality. Although the thesis was honoured by its publication in the prestigious *Annales Scientifiques de l'École Normale Supérieure*, Bachelier's career was far from smooth. He eventually obtained a permanent academic position at the university of Besançon.

Poincaré was not only an unusually gifted scientist; in his interest and choice of topics he could be unconventional. He supported and supervised Bachelier's work on financial mathematics. In addition, his student A. Quiquet, who edited Poincaré's lecture notes on probability (see [Poincaré 1890a]), worked on actuarial and statistical economic problems. In March 1906, Mittag-Leffler asked Poincaré to contact Quiquet for information regarding the founding of a Swedish actuarial society. Not surprisingly, Mittag-Leffler became president of this society [Poincaré 1999, letter 227]. Poincaré's answer a few weeks later contained the necessary information.

A second doctoral student associated with Poincaré was the Serbian Mihailo Petrović (1868–1943), who wrote a thesis on differential equations in 1894; apart from Poincaré, the mathematicians Hermite, Picard, and Painlevé played a part in the decision about its content. Petrović (see Figure 4.6) returned to Belgrade and became an influential scientist in Serbia.

Fig. 4.6 Mihailo Petrović
(1868–1943) wrote a
dissertation on differential
equations

The Romanian Dimitrie Pompeiu (1873–1954), see Figure 4.7, received his doctorate in 1905 for his work on complex function theory and later became a leading mathematician in Romania.

4.6   A French–English Controversy of Styles

In the lecture notes [Poincaré 1890a, nr. 2] on electricity and optics, *Electricité et optique*, the introduction takes a firm distance from Maxwell's style of developing theoretical physics. It begins thus:

> The first time a French reader opens the book by Maxwell, first a feeling of dejection and often even of defiance mixes itself with admiration. Only after long activity and at the cost of much effort does this feeling disappear. Some eminent spirits will always keep it.
>
> Why is it that the ideas of the English scientist have so much difficulty in feeling at home with us? This is without doubt because the education of the majority of the enlightened French gives them a taste for precision and logic before anything else.
>
> The old theories of mathematical physics gave us in this respect complete satisfaction. All our masters, from Laplace to Cauchy, have proceeded in the same way. Starting from clearly formulated hypotheses, they derived from these the consequences with mathematical rigour, and after that they have compared them with experiments. It seems that they wanted to give each of these branches of physics the same precision as celestial mechanics.

After deploring the lack of logical consistency and abstraction in Maxwell's writings, Poincaré states, "Maxwell does not give a mechanical explanation of electricity and magnetism; he restricts himself to demonstrating that such an explanation is possible." Poincaré points out that Maxwell's constructions are

Fig. 4.7 Dimitrie Pompeiu
(1873–1954) wrote a thesis
on complex function theory

preliminary and independent of each other. It is unclear what the relations between the hypotheses are. Some of the results are contradictory. In a solid theory of electricity, magnetism, and optics, one should be able to identify observable variables and to formulate Lagrangian equations of motion of the relevant quantities. Such a theory, systematically developed and tested against experience, would be considered an explanation of the natural phenomena.

Interestingly, this English style of mathematical physics persisted until far into the twentieth century. It also influenced the American scientists until, under the influence of continental European immigrants during the 1930s, a rationalization of theoretical physics in the United States took place. The strong internationalization of science in the second half of the twentieth century diminished the differences in scientific styles enormously, but the differences have not completely disappeared.

4.7   Relativity: The New Mechanics

The main priority controversy regarding the new mechanics, replacing Newtonian classical mechanics by relativity, is over special relativity, with prominent candidates Einstein, Lorentz, and Poincaré. The relativity of motion itself had already been studied and formulated by Galileo and Huygens, but still assuming the existence of an absolute reference frame for motion. Its perspective changed drastically following experiments conducted around 1900 showing that the velocity of light is independent of the inertial system chosen by the observer. Hendrik Lorentz (1853–1928) used this constancy of the velocity of light in each inertial system as the basis of his mechanics. It is the maximum velocity that can be observed, and to allow for this, he made the brilliant assumption that the size and mass of a body are dependent on the velocity in a given inertial system. His formula for the so-called Lorentz contraction gives this relation explicitly. In addition, Lorentz introduced the fundamental concept of local time, which means time as dependent on position and velocity in a given inertial system. So, like size and mass, time has no absolute meaning; there is no absolute reference frame for motion.

Poincaré noted already in 1900 that radiation could be considered a fictitious fluid with an equivalent mass; see also the discussion of these ideas in [Poincaré 1908a] and Section 11.3. He derived this interpretation from Lorentz's "theory of electrons," which incorporated Maxwell's radiation pressure. It is, of course, remarkable that Poincaré, who was always correct and even generous in citing people, did not mention Einstein in his 1909 lecture (see Section 11.4). This illustrates the fact that many physicists and mathematicians of that time considered Lorentz the prominent contributor to the theory of the new mechanics, now called special relativity. It is typical that still in 1913, Darboux wrote [Darboux 1913] that Poincaré discussed "the mechanics of Lorentz." There are indications, however, that Lorentz considered his observations provisional hypotheses, whereas Einstein presented a complete and new vision of physical reality, certainly by 1916, when he formulated the theory of general relativity. In 1927, Lorentz [Lorentz 1928] formulated the priority question at a conference as follows:

> I considered my time transformation only as a heuristic working hypothesis. So the theory of relativity is really solely Einstein's work. And there can be no doubt that he would have conceived it even if the work of all his predecessors in the theory of this field had not been done at all. His work is in this respect independent of the previous theories.

Perhaps this was overly generous. If for "heuristic" one reads "convenient," then most elements of special relativity were present in Lorentz's mechanics; later, Einstein extended the new mechanics to general relativity.

Poincaré and Lorentz show an ambivalence when mentioning the ether as a matter of fact in their writings. They seem to be reluctant to ignore it or to do without it; for Poincaré's mention of the ether, see, for instance, the essays "Theories of Modern Physics" in [Poincaré 1902] and "Science and Reality" in [Poincaré 1905b]. In his lecture [Lorentz 1915], Lorentz put it like this:

> Why can we not speak of the ether instead of vacuum? Space and time are not symmetric; a material point can at different times be at the same spot, but not in different places at the same time.

In [Borel 1914, note III], Borel notes a similar ambiguity in Poincaré's discussion of the "relativity of space." According to Borel:

> So it is maybe useful to clarify that his ideas on the relativity of space are ideas of the metaphysician and not of the scientist: the encompassing scientific apparatus adds nothing to the metaphysical doubt of the existence of exterior objects.

Still, in a discussion of priorities, the fundamental contributions of Lorentz to the formulation of special relativity theory, for instance his transformation formulas and the concept of local time, should be recognized together with Poincaré's contributions, namely his formulation of the Lorentz group and the principle of relativity; see Section 11.3. In a way, Lorentz's 1927 formulation given above should be supplemented with his appreciation of Poincaré's "dynamics of the electron" papers, given in [Lorentz 1914, p. 298]:

> Poincaré, on the other hand, has obtained a perfect invariance of the equations of electrodynamics and he has formulated the "relativity postulate" in terms that he was the first to use.

In a 1915 lecture at the Royal Academy of Sciences in Amsterdam, Lorentz put it as follows [Lorentz 1915]:

> I could point out to you [if I had more time] how Poincaré in his study of the dynamics of the electron, about the same time as Einstein, formulated many ideas that are characteristic for his theory, and also formulated what he calls "le postulat de relativité."

In this respect, it is difficult to understand why Einstein, when describing the development of relativity in 1949 [Einstein 1950], mentions many scientists, in particular Lorentz, but omits Poincaré.

4.8 Social Involvement

Even when he was very young, Henri Poincaré was interested in what happened in his town, in his country, and in the world. This is clear not only from his letters and conversations with friends; it appears also in his lectures and writings. In 1886, he assisted his father, Léon, in organizing a congress in Nancy on the progress of science. Later, he wrote popular articles on science in the *Revue Générale des Sciences*. His philosophical essays, most of which have been published in four books (see Chapter 6), are concerned with psychology, education, the foundations of mathematics, and the natural sciences. The topic of politics is not avoided.

In [Poincaré 1911], Poincaré discusses the relation between the sciences and the humanities, and here education naturally enters the picture. In 1912, he visited Vienna at the request of the "friends of the gymnasium" to talk about the future of the gymnasium in education. Here is a citation from [Poincaré 1911]: "Science has wonderful applications, but science that keeps its eye on applications only will not be science; it will be no more than a kitchen." Also in 1912, Poincaré gave a lecture on accepting the differences between people and the imperative to avoid hate between various social groups (see Chapter 12). Paul Appell [Appell 1925a] wrote how in 1904, the periodical *La Revue Bleue* sought the participation of scientists interested in politics. To a request of the editors, Poincaré answered:

> You are asking me whether scientists with political interest should fight or support the government. Well, this time I have to excuse myself; everybody will have to choose according to his conscience. I think that not everybody will cast the same vote, and I see no reason to complain about this. If scientists take part in politics, they should take part in all parties, and it is indeed necessary that they be present in the strongest party. Science needs money, and it should not be such that the people with power can say, science, that is the enemy.

In 1904, in the same periodical, *La Revue Bleue*, he gave his opinion on proportional representation in politics. Electoral systems were often discussed in those times, for instance by Joseph Bertrand in the context of probability theory. In [Poincaré 1913], Poincaré discussed the relation between ethics and science, the contrast between religious morality and scientific morality, which involves the emotional psychology of the scientist.

It is clear that Poincaré was often outspoken in his opinions, but he did not want to become anyone's tool. Of course, when he became famous, such attempts were made. He expressed himself very carefully about social questions without becoming vague. This can be seen, for instance, in his remarkable 1912 lecture "A plea for tolerance in society" (Chapter 12) for the Ligue française d'éducation morale.

It can also be observed in the question of the Dreyfus affair, which threw France into a social crisis around 1900. Because of the enormous consequences for the development of France and even its influence in Europe as a whole, we describe these events in more detail.

The Dreyfus Affair

At the end of the nineteenth century, a major political and social crisis held France in its grip: the affair of the trial and conviction of the army captain Alfred Dreyfus. A minor but conclusive part in this tragedy was played by science, which was used to obtain "proofs" of Dreyfus's guilt in his first trial in 1894. When a retrial was granted, a commission of three scientists, Gaston Darboux, permanent secretary of the Academy of Sciences; Henri Poincaré, president of the Academy; and Paul Appell, member of the Academy and dean of the Faculty of Sciences at the Sorbonne, was asked to report on the scientific reasoning that was in fact a crucial part of the accusations. The trial and retrial of the captain was in several ways unique in modern history because of the shock waves of emotion it sent through society; the discussions, articles, and books it evoked; and the many consequences for individuals, involving loss of jobs, imprisonment, even death by murder or suicide in a number of cases. Even today, the discussion of the affair means trouble in certain French army circles. To appreciate the work of the scientific committee, we review the main events of the affair (see also [Birnbaum 1994]).

The Start of the Affair

In December 1894, a young Jewish officer in the French army, Alfred Dreyfus, was found guilty of high treason. He was discharged from the army, his sabre broken in public, and was exiled for life to Devil's Island, a prison island near French Guiana. Four years later, the novelist and pamphleteer Émile Zola wrote his famous "J'accuse," in which he proclaimed the innocence of Dreyfus while accusing the military establishment of a crude violation of the rules of justice. This article produced a tremendous uproar in France, splitting the nation into two factions: the Dreyfusards, who called for "justice for Dreyfus," and the anti-Dreyfusards, who defended the actions of the establishment, in particular the army, while sometimes producing frightening outbursts of anti-Semitism.

At that time, relations between France and Germany were strained. During the Franco-Prussian War of 1870–1871, the Prussian army occupied part of France. The peace treaty signed in 1871 was unfavourable for the French, and a large part of eastern France was lost to Germany. Because of his excellent performance at the military school and his subsequent career in the army, Captain Dreyfus became attached to the General Staff in 1893. This was not as simple as it looks now. The French officer class was dominated by officers of noble birth who were royalist and Catholic. When Dreyfus was being considered for such a position, a general on the selection committee removed Dreyfus's name from the list of officers under consideration because he was Jewish. An official protest by Dreyfus—"Is a Jewish officer not able to serve his country as well as anybody else?"—convinced the committee that he was fit for the position.

In 1894, the French found among the belongings of a German officer a paper with classified information about the French military. The information must have come from the General Staff office, where, it was concluded, a spy was active. According to the counterintelligence of the French army, this could be only the Jewish officer working there, and Dreyfus was arrested immediately. In the preparation for the trial, the head of the research section of the *Préfecture* (the bureau of the prosecuting attorney), Alphonse Bertillon, played a crucial role. Bertillon dabbled in graphology and had some very weird ideas. The fact that the handwriting in the incriminating document did not resemble Dreyfus's was identified by Bertillon as *autoforgerie*. According to his report, the fact that he had disguised his handwriting was yet stronger proof that Dreyfus was a very accomplished spy.

After the trial, while Dreyfus was suffering on Devil's Island, his brother, Mathieu, was very active. He approached political and legal authorities to plead for a review of the process and to obtain new information to prove the innocence of his brother. He uncovered some remarkable legal irregularities, and the press again became interested in the case. Also, the handwriting on the document was identified as that of another officer of the General Staff, Ferdinand Walsin Esterhazy. It turned out that Esterhazy had been selling information about the French Army to the Germans, and at the end of 1898 he hurriedly left France.

In the meantime, the reaction to the evidence of Dreyfus's innocence was furious. Anti-Dreyfus nationalistic mass demonstrations took place in almost every French town. Around 10,000 people demonstrated in Paris, where anti-Dreyfus and anti-Semitic slogans were prominent. Most of the clergy, religious orders, and even the Vatican supported the anti-Dreyfusards. Interestingly, references to these events could be found even after the Second World War, when the propaganda of the Nazi-supporting Vichy regime was reviewed.

The Scientific Committee

By judicial order, the evidence used in the first trial was reviewed by the committee of three scientists from the Academy of Sciences consisting of Darboux, Poincaré, and Appell. In itself, the choice of such a committee was no guarantee of an impartial assessment. The Academy was and is clearly part of the establishment and as such not only a scientific but also a political body. When, for instance, Einstein visited Paris in 1922 to give some lectures, there were fears of nationalistic demonstrations. Einstein was a German! A lecture for the Academy of Sciences was organized and then cancelled when thirty members of the Academy, including its president, announced that they would boycott the event.

However, the committee to review the evidence in the Dreyfus trial proved to be impartial and in complete agreement about all details in their report. When it became known that Dreyfus had been condemned on evidence that was not known to the defence, Poincaré said to Appell [Appell 1925a], "The enormity of the accusation has probably destroyed the critical sense of the judges." The committee was asked to consider the graphological analysis of Bertillon that was tied in with a probability calculation. In his letter to the court, Poincaré wrote as chairman of the committee, "There is nothing scientific in this evidence and I cannot understand your uneasiness. I do not know whether the accused will be found guilty, but if he is, it will be on other proofs. It is impossible that such an argumentation would be seriously considered by scientifically educated people without prejudice." Appell noted that during the whole proceeding, Poincaré showed a certain impatience because of the triviality of the questions put to the committee. Still, he considered the questions conscientiously and gave them his full attention.

According to Appell [Appell 1925a], Poincaré's attitude during these proceedings was typical of his views on public morals. This is expressed in a more general context in a lecture given in 1912, in which he commented on the differences between people and discussed hate in society; see Chapter 12.

Retrial, Pardon, and Rehabilitation

In 1899, following a request by his wife, Lucie, Dreyfus obtained a review of the legal proceedings, and a retrial was held before the military court of Rennes. Against all logic and reason, Dreyfus was again found guilty, this time with a sentence of ten years' imprisonment. However, because of his weak health, Dreyfus asked the president of the republic, Émile François Loubet, to pardon him. The president granted the pardon that same year.

In 1903, the politician Jean Jaurès reopened the case. The officer in charge discovered even more fabricated evidence, and in 1906, a higher court determined in a third trial that the accusation against Dreyfus was without any foundation. He was reinstated in the army and awarded the Légion d'Honneur. His last words were, "I was only an artillery officer who by a tragic error was prevented from following his career."

Chapter 5
The Prize Competition of Oscar II

In the summer of 1885, an announcement of a prize competition appeared in several scientific journals. The announcement, from Gösta Mittag-Leffler, stated that King Oscar II of Sweden and Norway had decided to sponsor a scientific competition, with a prize to be awarded on January 21, 1889, his 60th birthday. The practical aspects of the competition were the responsibility of three committee members: chairman Gösta Mittag-Leffler (Stockholm), Karl Weierstrass (Berlin), and Charles Hermite (Paris). The prize would consist of a gold medal and the sum of 2500 kronor. The memoirs offered for the competition were to be submitted by June 1, 1888.

The participants could choose from four topics. The first, which was taken up by Henri Poincaré, was formulated as follows (see [Barrow-Green 1997]):

A system being given of a number whatever of particles attracting one another mutually according to Newton's law, it is proposed, on the assumption that there never takes place an impact of two particles, to expand the coordinates of each particle in a series proceeding according to some known functions of time and converging uniformly for any space of time.

It seems that this problem, the solution of which will considerably enlarge our knowledge with regard to the system of the universe, might be solved by means of the analytical resources at our present disposition; this may at least be fairly supposed, because shortly before his death, Lejeune-Dirichlet communicated to a friend of his, a mathematician, that he had discovered a method of integrating the differential equations of mechanics, and that he had succeeded, by applying this method, in demonstrating the stability of our planetary system in an absolutely strict manner. Unfortunately, we know nothing about this method except that the starting point for its discovery seems to have been the theory of infinitely small oscillations. It may, however, be supposed almost with certainty that this method was not based on long and complicated calculations but on the development of a simple fundamental idea, which one may reasonably hope to find again by means of earnest and persevering study.

However, in case no one should succeed in solving the proposed problem within the period of the competition, the prize might be awarded to a work in which some other problem of mechanics is treated in the indicated manner and completely solved.

In retrospect, this trust in the correctness of Dirichlet's statement seems naive. One is reminded of Fermat's last theorem, of which he wrote that the margin of the book in which he formulated it was too small to contain the proof. It might be supposed that

F. Verhulst, *Henri Poincaré: Impatient Genius*, DOI 10.1007/978-1-4614-2407-9_5,
© Springer Science+Business Media New York 2012

Dirichlet or Fermat was in possession of such a method or proof, but it seems highly improbable. Also, to allot three years to solving a problem that is still completely unsolved today seems now rather ambitious.

In this chapter we will describe the outcome of the competition; a monograph with many more details and references is [Barrow-Green 1997].

5.1   Comments by Kronecker and Start of the Competition

The announcement of the prize competition in 1885 led to an angry letter to Mittag-Leffler from Leopold Kronecker (1823–1891). It is an understatement to say that Kronecker was no friend of Weierstrass (see Figure 5.3), and the composition of the prize committee clearly caused him irritation. Apart from formal complaints about the composition of the committee and the way the competition had been announced, Kronecker had one material objection: Problem four posed by the committee was concerned with algebraic questions regarding Fuchsian functions. Kronecker claimed that he had proved that the results asked for in the announcement could not be achieved. He threatened to write to the king about this point. Mittag-Leffler pleaded ignorance, and fortunately, Kronecker let the matter rest.

In the same year, 1885, Kronecker raised in a letter to Mittag-Leffler the question of the formulation of problem one. He claimed that he was the "friend" mentioned by Dirichlet, and accordingly, he was the only person who could describe what Dirichlet had communicated to him. Three years later, in 1888, Kronecker went public with this information, adding that Dirichlet had been misquoted. This was probably mainly an attack on Weierstrass, but the committee as a whole was expected to react. Hermite (see Figure 5.1) did not want to be involved in "this

Fig. 5.1 Charles Hermite, member of the prize committee

German affair," while Mittag-Leffler and Weierstrass concluded that apart from omitting the name of the "friend," there had been no incorrect formulation. They decided to ignore Kronecker's attack.

The identity of the entrants was supposed to be secret, but Poincaré told Hermite and Mittag-Leffler that he intended to submit a memoir on problem one. By the close of the competition, June 1888, twelve contributions had been received, five of them dealing with problem one.

5.2   Activity and Conclusions of the Committee

Mittag-Leffler began by giving the twelve memoirs to a younger colleague, Lars Edvard Phragmén (1863–1937), see Figure 5.2, with the assignment to make a preselection. By early in the summer of 1888 he wrote to Weierstrass and Hermite that only three contributions were of real interest, two from Paris (Poincaré's and Appell's) and one from Heidelberg. The committee looked closely at these three contributions and soon reached the unanimous decision that Poincaré should be awarded the prize. Appell would receive an honourable mention. Paul Appell, by the way, had chosen a topic to his own taste, the expansion of abelian functions.

The next stage was much more difficult for the committee; there was to be a public appraisal of the contributions, and the winning memoir published in the *Acta Mathematica*. As usual with Poincaré's manuscripts, there were large gaps in the proofs and many intuitive steps. One may also safely say that the full content of the prize memoir and the validity of its statements were not completely understood by the committee members. How could they then award the prize? Their position is summarized by Hermite in a letter to Mittag-Leffler [Hermite 1985] (translation [Barrow-Green 1997]):

Fig. 5.2 Lars Edvard Phragmén, editor of the prize memoir

Fig. 5.3 Karl Weierstrass, member of the prize committee

> Poincaré's memoir is of such rare depth and power of invention, it will certainly open a new scientific era from the point of view of analysis and its consequences for astronomy. But greatly extended explanations will be necessary, and at the moment I am asking the distinguished author to enlighten me on several points.

Hermite could ask the author, but Poincaré remained vague in his answers regarding statements he considered self-evident. Mittag-Leffler, however, took the unusual step of asking Poincaré to add explanations to the memoir, in this way mixing his position as chairman of the prize committee and his position as editor of the *Acta*. Poincaré's memoir would have occupied 158 pages of the *Acta*, but he added an additional 93 pages with notes.

King Oscar II announced on his birthday, January 21, 1889, that Poincaré had won the prize competition and that Appell had been awarded an honourable mention. The French newspapers made much of this victory for France, and both scientists were awarded the Légion d'Honneur by the French government. The winning memoir together with the memoir by Appell was to be published in the *Acta Mathematica* of October 1889, but as we shall see, a shocking development intervened.

5.3   A Blessing in Disguise

The young mathematician Phragmén was given the task of editing Poincaré's memoir for the *Acta Mathematica*. Very soon, in the summer of 1889, he found in part of the memoir statements and conclusions that were not clear to him. Mittag-Leffler did not realize how serious one of those points was, and he asked for additional clarification. This made Poincaré look again at his work, and he began to have doubts about certain convergence arguments. After several months of agonizing, he wrote to Mittag-Leffler in December 1889 [Poincaré 2012] (translation [Barrow-Green 1997]) that substantial changes had to be made:

> I have written this morning to Phragmén to tell him of an error I have made, and doubtless he has shown you my letter. But the consequences of this error are more serious than I first thought. It is not true that the asymptotic surfaces are closed, at least in the sense that I originally intended. What is true is that if both sides of this surface are considered (which I still believe are connected to each other), they intersect along an infinite number of asymptotic trajectories (and moreover, their distance becomes infinitely small of order greater than μ^p, however great the order of p).
>
> I had thought that all these asymptotic surfaces, having moved away from a closed curve representing a periodic solution, would then asymptotically approach the same closed curve. What is true is that there is an infinity of them that enjoy this property.
>
> I will not conceal from you the distress this discovery has caused me. In the first place, I do not know whether you will still think that the results that remain, namely the existence of periodic solutions, the asymptotic solutions, the theory of characteristic exponents, the nonexistence of single-valued integrals, and the divergence of Lindstedt's series, deserve the great award you have given them.

What seemed a catastrophe at the time turned out to make the prize-winning memoir in its final form even more important. Suddenly, it became the first paper touching upon the subject of nonintegrability and chaos of dynamical systems. It was, however, so far ahead of its time that it took till around 1960 for many scientists to become aware of the importance of this work. The remark by Poincaré that "there is an infinity [of asymptotic surfaces] that enjoy this property" anticipates the more general KAM theorem, formulated and proved around 1960 by Kolmogorov, Arnold, and Moser. It should be stressed that in trying to answer Phragmén, Poincaré found the error himself, and that apart from the erroneous section, the memoir contained many fundamental and beautiful results, in itself enough for the prize to be awarded to him several times over.

In the meantime, Mittag-Leffler was saddled with the responsibility of avoiding a scandal and dealing with the practical problems that had arisen. He was a skilful operator, but this case was not easy, since several scientists, including Kronecker, Gyldén, Lindstedt, and the other members of the committee, were very interested in the prize memoir and of a very critical disposition. The king and the public would not be informed.

The *Acta* publication with the memoirs of Poincaré and Appell had already been printed but not yet distributed except for about twenty copies. Under several pretexts, Mittag-Leffler asked for the advance copies to be returned. A new version had to be printed at the cost of 3500 kronor, a formidable sum for a print run. Poincaré was asked to pay the costs, which he did without comment. Note that the annual salary of a Swedish professor at that time was around 7000 kronor, while the prize money was 2500 kronor. The other two committee members were informed step by step, which was painful to them, since they had given their approval of the first version. Weierstrass, who had originally formulated problem one of the competition, felt especially bad about it.

And so a scandal was avoided by the discretion of Mittag-Leffler. Almost all the printed copies of the first version of the memoir were destroyed, while

Fig. 5.4 Henri Poincaré in
1889, age 35, when he won
the prize awarded by King
Oscar II of Sweden

Mittag-Leffler kept as much information about the affair to himself as possible.
When finally the memoir [Poincaré 1890b] appeared in 1890, there were very
few scientists who understood what Poincaré was getting at. There were critical
comments by astronomers such as Hugo Gyldén (1841–1896) and Anders Lindstedt
(1854–1939), but these were relatively easy to refute, since the series expansions
they used were ingenious but entirely formal, i.e., without a proof of their validity.
Also, they did not touch upon the intricate dynamics described by Poincaré. One of
the first scientists to grasp the importance of Poincaré's results seems to have been
a young German mathematician, Hermann Minkowski (1864–1909).

5.4   The Prize Memoir

Most of the material in the prize memoir [Poincaré 1890b] is included and extended
in the three volumes of the *Mécanique Céleste*, so we leave this discussion to Section
9.3. We will discuss here the erroneous section of the first version of the memoir,
which shows a very natural line of thinking that since then has been repeated many
times by scientists who have not been aware that the matter has been settled in
a very general way. A loose way of formulating it is that in general, nonlinear
conservative systems are nonintegrable, except for isolated cases. It is misleading
that a number of real-life models, such as the gravitational two-body problem, are
integrable. This led scientists to believe that the solution of differential equations
in terms of integrals involving elementary or special functions was only a matter of
ingenuity and diligence. They did not yet realize that there could be a fundamental
obstruction to such solutions.

As mentioned, a famous example of integrability is the gravitational two-body problem, for which all the solutions can be nicely classified as ellipses, hyperbolas, or parabolas located on smooth manifolds completely filling phase space. With this knowledge, it was then quite natural to expect that a small extension of the problem, for instance by adding a third small mass, could be handled in a similar way. This was the line of thinking that Poincaré took in his first version of the prize memoir, which addressed the problem of describing the positions of n bodies moving in their mutual gravitational field. Poincaré considered the case $n = 3$, but reduced the problem even more, by assuming that of the three masses, two were significant, with the third being so small that it does not affect the motion of the other two. This is called the *restricted three-body problem*. It means that for the two larger bodies we can use the solutions of the two-body problem and that we attempt to describe the motion of the body with negligible mass as it moves in the gravitational field of the two larger bodies. Such a situation models, for instance, the motion of a spacecraft or asteroid in the field of the Sun and a planet such as Jupiter or Earth. It is noteworthy that even this restricted three-body problem can be considered completely unsolved even today. For although a large number of special solutions have been found and although we can obtain special numerical solutions for given initial conditions, we have no general picture of the behaviour of the orbits in 6-dimensional phase space.

Poincaré put even further conditions on the restricted three-body system by assuming that the three masses were moving in a plane in physical space; this is called the *planar restricted three-body problem*. In part of his contribution, he reduced the model even more: the two larger masses moved in circular orbits; this is called the *planar, circular, restricted three-body problem*. It leads to a problem with two degrees of freedom, described by four first-order differential equations with periodic coefficients. It should be remarked that the methods that Poincaré developed for this problem in the memoir were very innovative and can be used for more general problems outside celestial mechanics and for arbitrary dimensions.

The problem with four first-order differential equations that Poincaré considered contained equilibrium solutions and periodic solutions. If they are unstable, some solutions will be attracted to (say) a periodic solution and some will be repelled. The set of attracted solutions forms a manifold (the stable manifold), and the set of repelled solutions also forms a manifold (the unstable manifold); Poincaré called these manifolds "asymptotic surfaces." For the solutions on these surfaces he used convergent series expansions, and in this way he could follow the unstable manifold as it moved away from a periodic solution with increasing time, and the stable manifold by considering negative time. His first incorrect assumption was that the continuations of stable and unstable manifolds would lead to smooth gluing of both to produce a single manifold. Today, this is called a *homoclinic manifold*; Poincaré used this term later in his books on celestial mechanics [Poincaré 1892], as well as the term "asymptotic surface." The existence of a family of such homoclinic manifolds filling phase space (asymptotic surfaces in the terminology of the memoir) would correspond to the existence of an integral invariant of the dynamical system.

Only following Phragmén's queries did Poincaré discover that this smooth gluing is generally impossible, implying that another integral invariant for his reduced three-body problem does not exist. He also realized that for this problem, the intersections of stable and unstable manifolds take place on a smaller than algebraic scale. This means that with the presence of a small parameter μ, for instance scaling one of the masses, the transversality angle of stable and unstable manifolds is smaller than μ^n for arbitrarily large n. It requires remarkable technical skill to prove this result, which we know now to be valid for general Hamiltonian systems exhibiting two degrees of freedom. At the same time, the intersection of stable and unstable manifolds opens up the possibility of irregular dynamics, which we call "chaos." It should be mentioned that these nonintegrability results hold for models that are more complicated than those considered by Poincaré, and such results become general and even more prominent for higher dimensions.

A visionary description of the ensuing chaotic dynamics that was far ahead of its time can be found in the last chapter of [Poincaré 1892, Vol. 3]; see also Section 9.3 of this book.

Chapter 6
Philosophy and Essays

> *Geological history shows us that life is only a short episode*
> *between two eternities of death and that even within this*
> *episode, conscious thinking has not lasted and will not last more*
> *than a moment. Thinking is a light ray only in the middle of a*
> *long night. But it is this light ray that counts.*
> —Henri Poincaré, *La Valeur de la Science*

A number of Henri Poincaré's essays are usually classified as "philosophical." Most of them have been collected in six books. The first five books do not mention the provenance of the original papers, which appeared in various periodicals, such as *Revue de Métaphysique et de Morale*, often with mathematical details and references that were left out in the book versions. The omission of sources made the books accessible to a wide public, and there were probably also marketing considerations, since the absence of sources suggested a greater originality of the writings. The last collection, *Scientific Opportunism*, published in 2002, makes up for this omission by including a list of sources and a description of the background of the five books that appeared in the years up to 1913, which were published by Ernest Flammarion, whose brother Camille was an amateur astronomer and prolific popularizer. Camille Flammarion wrote 31 popular books on science, all published by Flammarion. However, the Flammarion brothers were not scientific insiders, and so Ernest Flammarion was happy to welcome a proposal from Gustave Le Bon (1841–1931), see Figure 6.1, to establish a *Bibliothèque de Philosophie Scientifique*. Le Bon was not a professional scientist, but he was an intellectual with wide-ranging knowledge and connections. The *Bibliothèque de Philosophie Scientifique*, which included items 1, 2, 3, and 5 in the list below, would bring a fortune to both Flammarion and Le Bon. The six collections of essays along with their dates of publication are given in the following list:

1. *La Science et l'Hypothèse* (1902) [Poincaré 1902]
2. *La Valeur de la Science* (1905) [Poincaré 1905b]
3. *Science et Méthode* (1908) [Poincaré 1908a]

F. Verhulst, *Henri Poincaré: Impatient Genius*, DOI 10.1007/978-1-4614-2407-9_6,
© Springer Science+Business Media New York 2012

Fig. 6.1 Gustave Le Bon
(1841–1931), French
intellectual with many
contacts in science,
psychology, and sociology;
editor of Éditions
Flammarion

4. *Savants et Écrivains* (1910) [Poincaré 1910]
5. *Dernières Pensées* (1913) [Poincaré 1913]
6. *Scientific Opportunism: An Anthology* (2002) [Poincaré 2002]

The first book of the series was Poincaré's *La Science et l'Hypothèse*, which sold 5000 copies in the first six months, and 21,000 altogether by 1914. Around 1900, Poincaré had become famous in scientific circles, but after the publication of *La Science et l'Hypothèse*, he became a national celebrity. *Savants et Écrivains* contains biographical sketches and observations about prominent people; it was published by Flammarion outside the series and was not particularly successful. It was not reprinted.

To what extent are these writings philosophical? René Thom seems to think that they are not, finding Poincaré too much of a mathematician to be a philosopher [Thom 1987, p. 72]. Moreover, he lacks the ontological point of view that according to him characterizes a great philosopher. It is a point of view that is difficult to understand considering Poincaré's main concern: the value and meaning of scientific activity. If a criterion for philosophy is that it be a "serious reflexion on scientific truth and human behaviour," then Poincaré's essays are certainly within the scope of philosophy, as we will see in his discussion of conventionalism in mathematics and physics.

Controversy among philosophers over Poincaré's philosophical writings began very soon. Bertrand Russell wrote in his preface to the English translation of *Science et Méthode* by Francis Maitland [Russell 1914]:

> Readers of the following pages will not be surprised to learn that [Poincaré's] criticisms of mathematical logic do not appear to me to be among the best parts of his work. He was already an old man when he became aware of the existence of this subject, and he was led by certain indiscreet advocates to suppose it in some way opposed to those quick flashes of insight in mathematical discovery which he has so admirably described.

Poincaré was, however, only in his forties when he started writing and lecturing about mathematical reasoning, for instance at the international mathematical congress in 1900, and around fifty when the controversies with the logicians began; see Section 6.2. The points that Poincaré raised regarding mathematical logic were not solved during his lifetime.

A problem for philosophers, in particular for those of the logical persuasion, is that Poincaré does not fit well into the classical discussions and problem formulations, whereas his writings doubtless play a part in philosophical discussions of the foundations of science. His starting points were always his own mathematical thinking and physical facts, which led him to theorize about truth. One should keep in mind that the beginning of the twentieth century was a period of grand logical designs and exaggerated claims. The mathematical-logical program of the period did little justice to the needs of the working mathematician. A reduction of mathematics to logical steps was correctly rejected by Poincaré, since it would deprive mathematics of the living source of intuition ("Anschauung"). Poincaré's essays on logic in *Science et Méthode* are more in the direction of brilliant pamphlets exposing the weak points in the program of logicism than scholarly expositions on the topic; the logicians were left with the burden of refuting his observations.

In what follows we will briefly describe some of the recurring themes of the essays. As mentioned above, the essays in the books were taken from articles in periodicals or books, often with modifications. Most of the formulas were suppressed to make the essays more attractive to a general public; indeed, whole sections were omitted and others added. The sources and modifications have been given by Laurent Rollet in [Poincaré 2002]. It should also be noted that the collected works [Poincaré 1916] are not quite complete with regard to the philosophical and popular writings. The anthology [Poincaré 2002] is also of interest for its description of the relation between Henri Poincaré, the publisher Flammarion, and the editor Gustave Le Bon, who maintained an extensive network of artists, scientists, and politicians. Le Bon organized dinners and other social gatherings on a regular basis with Henri and Raymond Poincaré, Paul Painlevé, Camille Saint-Saëns, Paul Valéry, and many other prominent figures.

6.1 The Last Collection: Scientific Opportunism

The history of the anthology *Scientific Opportunism* [Poincaré 2002] is remarkable. A young teacher, Louis Rougier, approached Gustave Le Bon with the proposal to publish posthumously a fifth volume of Poincaré's philosophical essays. Le Bon was interested, and around 1919, he wrote to Poincaré's widow, Louise, asking permission to publish this new volume. Her brother-in-law Émile Boutroux and his son Pierre had already considered the project, and she had been told that they approved. The title would be *L'opportunisme scientifique*.

Louise Poincaré was hesitant because her brother-in-law had given only superficial consideration to the project and perhaps also because she was suspicious of the commercial interests of Le Bon and the publisher. She asked the advice of her eldest daughter, Jeanne, and her daughter's husband, Léon Daum; Daum was a prominent engineer and what we would today call a captain of industry. After looking at the material, Daum concluded that the proposed collection of articles contained no new ideas but only small additions and variations of older work. He proposed that some of the articles be included in a new edition of the *Dernières Pensées* [Poincaré 1913], and this was done in the edition of 1926, to which four articles were added.

The present anthology [Poincaré 2002] contains the original articles proposed by Rougier. They deal with the foundations of geometry, celestial mechanics, and various other topics in science. A lecture for the University of Brussels in 1909 [Poincaré 1913, Poincaré 2002] presents explicitly Poincaré's views on the relation between science and religion. This was a hot topic in France around 1900, when the Roman Catholic Church maintained extremely conservative positions on political, social, and scientific questions of the time.

After stating that lawyers are concerned not with finding the truth but convincing a judge, Poincaré notes that it is of great importance to exclude all preconceived opinions. He writes in [Poincaré 2002, p. 141]:

> Do not understand this in the sense that I want to prohibit science to religious people, in particular to Roman Catholics. God forbid! I would not be so stupid as to deprive humanity of the services of a Pasteur. There are those who forget their beliefs when they enter a laboratory; as soon as they have their work clothes on, they put a bold face on truth and they have as much critical spirit as anyone else.

This suggests that those who are both religious and good scientists manage to live in two separate worlds at the same time. The observation fits in with the opinion of Poincaré's friend Paul Appell [Appell 1925a, p. 79], which can be phrased thus:

> Religious truth varies on Earth with latitude and longitude, but science is one. Religion is something for the individual conscience; scientific truth is the same for everybody and all conscious beings in the universe. Science progresses by small successive approximations; it will never show the final truth of everything.

It is not only dogmatic thinking in religion that is a danger to the search for scientific truth. Any dogmatism will impede scientific development according to Poincaré in [Poincaré 2002, p. 141]:

> The dogmas of the religions founded on revelation are not the only ones to fear. The impression Catholicism has made on the Western soul has been so deep that many recently liberated spirits have a nostalgic need for dependency and have been making an effort to rebuild churches. In this way, certain schools of positivism are nothing but a Catholicism without God. Auguste Comte himself dreamt of disciplining minds, and some of his disciples who exaggerated the idea of the master quickly became enemies of science when they obtained the upper hand. All exterior restraint is nothing but an obstruction to thought, and it would not be worth the effort to shatter the old one while accepting a new one.

In his writings, Poincaré was interested primarily in obtaining insight into physical reality and mathematical reality—two different things. This should not be taken as an overly restrictive program, for it also involved wide-ranging reflection on how the mind works, how mathematics should be taught, and how great scientists developed their ideas. His style of writing is in beautiful classical French, crystal clear and directed right to the heart of the matter. His writings have a freshness that derives from the actual experience of a creative mind in mathematics and physics, something most philosophers lack. It is understandable that an eminent scientist who could write so well would play a prominent role in public discourse.

The philosophical essays of Henri Poincaré are far from an attempt at systematic reflection. They can be seen as a critical assessment of scientific activity in its various forms. Setting apart the descriptions of scientists [Poincaré 1910], one can distinguish certain recurring themes in the collections. Regarding the essays on the foundations of mathematics and the natural sciences, the key words are *convention*, *hypothesis*, *objectivity*, and *intuition*. However, in different disciplines these notions can have different meanings.

6.2 The Foundations of Geometry and Mathematical Thinking

The topics of mathematical foundations and modes of reasoning can be found in all Poincaré's philosophical books. The original essays were written both before and after the appearance of Hilbert's *Grundlagen der Geometrie* [Hilbert 1899]. The contrast between the two great mathematicians can be found largely in the axiomatic approach of Hilbert and the more intuitionistic thinking of Poincaré; for a discussion of these issues, see [Eymar 1996, Sanzo 1996, Heinzmann 2010].

A primary notion is that the existence of an object in mathematics and the existence of a material object are very different things. A mathematical entity such as a point or a triangle exists if its definition does not imply a contradiction, either in itself, or within earlier accepted theorems [Poincaré 1902, p. 59]; see also [Poincaré 1908a]. So existence is boldly identified by Poincaré with noncontradiction. Surprisingly enough, Poincaré is in agreement on this point with the logicists and with Hilbert. This is one of the main points on which French preintuitionism differs from later intuitionism as represented by L. E. J. Brouwer.

The development of the theory of mathematical objects requires axioms, both explicitly formulated and implicitly used. However, in mathematics one is free to make choices about such axioms. Euclidean geometry, for instance, is based on a consistent system of axioms, but in the nineteenth century, Lobachevsky, Bolyai, and Riemann showed that by choosing another system of axioms, one could develop non-Euclidean geometries that are also consistent. Lobachevsky accomplished this by dropping only one of the Euclidean axioms, the notorious "fifth postulate": Given a point external to a given straight line, one can draw precisely one line through this point parallel to the given line.

The axioms (or hypotheses) selected by a mathematician are called "conventions" by Poincaré. One can drop a convention and adopt another one, provided that one shows that the corresponding geometry, or in general the corresponding mathematical theory, contains no contradictions, that is, that it is consistent in the mathematical sense. This view of mathematics is called *conventionalism*.

Fig. 6.2 Louis Couturat
(1868–1914), mathematical
logician, had views on the
foundations of mathematics
that differed from those of
Poincaré

Note that these observations are concerned with mathematical concepts like space and dimension. As we shall see, conventions and conventionalism play a very different role in the natural sciences. Instead of speaking of conventions of the natural sciences, it is better to avoid confusion with conventionalism in mathematics and to speak of "convenient hypotheses." Both in mathematics and in the natural sciences, "truth" is what we are looking for, but in mathematics, "truth," according to Poincaré, is determined by conventions. We have objective mathematical truth of statements and theory only *within* a given system of conventions.

A spirited and sometimes vehement discussion took place between Poincaré and the French logician Louis Couturat (1868–1914), see Figure 6.2, about the importance of logic in mathematical theory. Since the end of the nineteenth century, prominent mathematicians including Peano, Frege, and Hilbert had been involved in questions of the foundations of mathematics by building mathematics on the basis of a set of axioms and a set of rules for logical deduction.

Building up mathematics in this axiomatic setting should lead to consistent mathematical theories, without contradictions. Poincaré disagreed with this assertion and emphasized "intuition" as a fundamental ingredient in mathematical thinking in [Poincaré 1908a, Poincaré 1902, Poincaré 1913]. He stated, for example, that mathematical induction, one of the tools very often used in mathematics, could not be justified by logic alone; it needs *intuition*. Mathematicians always want to go from special examples to much more general statements. They want to show how special features are part of much more general characteristics. How can one then call mathematics deductive? [Poincaré 1902, Chapter 1]. Poincaré's arguments certainly made sense, and they stimulated thinking about the relation between mathematics and logic, especially outside France. Unfortunately, in France, Poincaré's authority caused a delay in important discussion of the role of mathematical logic.

Mathematical Induction

One of the basic tools of mathematical reasoning is called proof by induction. We illustrate this by a simple example. Suppose we consider the positive integers $1, 2, 3, \ldots, n, \ldots$, where n symbolizes an arbitrary positive integer. We would like to add these numbers in sequence up to some point. For instance, $1 + 2 = 3$ represents the sum of the first two positive integers, $1 + 2 + 3 = 6$ is the sum of the first three, and so on. Let us indicate the sum up to the nth number by the symbol S_n, where n therefore indicates the last number of the summation. So $S_2 = 1 + 2 = 3$, $S_3 = 1 + 2 + 3 = 6$, and so on. Carrying out this summation for a large number n is a lot of work, but fortunately, mathematicians have worked out a formula for this sum: for any positive integer n, we have

$$S_n = \frac{1}{2}n(n + 1).$$

How can we prove that the formula is correct? It is clearly correct for $n = 2$, as is easily checked from the formula $S_2 = \frac{1}{2}2(2 + 1) = 3$. In the same way, we can also check that it is correct for $n = 3$. Mathematical induction will now give us the correctness for arbitrary n. Suppose the formula is correct for some n (this is the case for $n = 2$ and $n = 3$). Based on this supposition, we will show that the formula must also be correct for $n + 1$. What, then, will be the sum S_{n+1}? We have

$$S_{n+1} = 1 + 2 + 3 + \cdots + n + (n + 1).$$

Because we have assumed that the formula for S_n is correct, we may write

$$S_{n+1} = \left[1 + 2 + 3 + \cdots + n\right] + (n + 1) = S_n + (n + 1)$$

and then replace S_n in the above equation by by $\frac{1}{2}n(n + 1)$, obtaining

$$S_{n+1} = \frac{1}{2}n(n + 1) + (n + 1) = (n + 1)\left(\frac{1}{2}n + 1\right) = \frac{1}{2}(n + 1)(n + 2).$$

So the formula holds for $n + 1$ on the assumption that it holds for n. We conclude that the formula holds for arbitrary n.

Mathematical induction works in the same way for an arbitrary statement about the positive integers. Suppose we wish to prove a proposition depending on the positive integers $1, 2, 3, \ldots, n, \ldots$. We ascertain that the proposition is correct for a certain value of n, for instance $n = 1$ or $n = 2$. We then assume that the proposition is correct for arbitrary n and show that from this assumption, it follows that the proposition is correct for $n + 1$. Since we tested the correctness for $n = 1$ or $n = 2$, the proposition must be correct for $n = 3$, and therefore for $n = 4$, and so on. That is, it must be correct for arbitrary n.

Regarding the foundations of mathematics and the handling of paradoxes in logic, both Bertrand Russell and Henri Poincaré were thinking about proper characteristics of *definitions and classifications*. Regarding "classification," think of a set whose elements are distinguished or classified by different characteristics; for instance, the set of integers has the subsets of even and of odd numbers. Poincaré calls a classification *predicative* [Poincaré 1913] if the classification is not changed by the introduction of new elements. We may classify, for instance, the set of the first hundred integers as consisting of a subset of numbers that are less than or equal to 10 and a subset of numbers that are greater than 10. Consider now the set of the first two hundred integers, which of course contains the first set. In considering this larger set, the classification of the first hundred integers does not change, and so the classification is predicative. A *nonpredicative* classification will change the classification of elements of a set when new elements are added. A definition is in fact a classification, so we require a definition to be predicative.

A famous article in [Poincaré 1908a] is concerned with *mathematical invention* (or discovery). The title Poincaré gave to the essay is "L'invention mathématique." The word *invention* in French can mean invention, discovery, or finding. The English language also exhibits this ambiguity, but with more stress on something newly invented. In the English translations of this essay, the choice is generally made for "discovery," implying that all of mathematics has a Platonic existence, waiting to be discovered. On the other hand, the French mathematician Jacques Hadamard (1865–1963) became interested in the subject and wrote an essay called "Essay sur la psychologie de l'invention dans le domaine mathématique." Hadamard supervised the English translation of his own book, which uses "invention" instead of "discovery" [Hadamard 1990], an important difference. Is it possible that the mathematicians who supervised the English translation of Poincaré's essays did not believe that part of mathematics is invention?

In a wonderful piece of introspection, Poincaré describes in the essay how sudden insight came to him in solutions of mathematical problems. He conjectures that the unconscious mind, stimulated by intense but seemingly fruitless exploration of a problem by the conscious mind, considers many mathematical combinations and makes a choice on the basis of aesthetics or economy. An example that he gives of such an occurrence concerns the Fuchsian functions. It is described in Section 3.3 of this book. In [Hadamard 1990], Hadamard extends these ideas. One must conclude that the use of the word "discovery" stands in contradiction to Poincaré's view of the role of conventions in mathematics. If one considers a fundamental problem in mathematics, a certain mathematical structure is selected in a process that is nearer to invention. After a long period of thought, a flash of insight gives the right approach or even the solution.

Remarkably enough, a kind of confirmation of these ideas came posthumously from Amédée Mannheim [Mannheim 1909]. In 1902, a journal for the study of teaching of mathematics, *L'Enseignement mathématique*, asked its readers to describe how they developed mathematics, how ideas emerged. Colonel Mannheim, who taught geometry at the École Polytechnique, wrote down some notes regarding these questions, but he did not send them to the journal, since "he did not want to talk in public about himself" [Mannheim 1909]. His notes, published posthumously in [Mannheim 1909], mention that there is no unique way to do mathematics, and for him, it was something that "went by itself" when he was walking, sitting in a noisy omnibus, or attending a concert. Mathematical invention came for him from the nearly unconscious activity of his brain, certainly not when he made a special effort while sitting at his desk.

On the topics of intuition and logic in mathematics, there is a very interesting essay, "L'intuition et la logique en mathématiques," in [Poincaré 1905b]. One can distinguish among mathematicians those who are mainly guided by step-by-step logic and those led by intuition, two different styles. One can also distinguish analysts from geometers, but it is not the content of mathematics that dictates the mathematical style. For instance, Felix Klein discusses geometric problems in complex function theory in a very intuitive way, but with the suggestion that the treatment is rigorous. On the other hand, Weierstrass develops analysis without drawing or referring to figures; he is a typical logician. Riemann develops his mathematics without drawing figures, evoking intuitively geometric images that are necessary for understanding his line of thinking.

Intuition is of great importance, but in most cases it gives no certainty. A curve, for instance, will have, according to intuition, at most points a tangent, but we know that there exist curves that are nowhere differentiable; hence they have no tangents. So we start intuitively and then add reasoning to make the proofs rigorous, a process that has been refined and improved in the course of scientific history. There are, however, different kinds of intuition. One is based on form and imagination, such as our (false) intuition on curves and tangents. This kind of intuition is appropriate to start with, but should be followed up by rigorous treatments. Mathematical induction needs another type of intuition; according to Poincaré, it cannot be validated by logical steps, but it is justified by intuition. A proof by induction can certainly be considered rigorous.

Logic and intuition each play a necessary part in mathematics. Logic is the instrument of proofs, while intuition is the instrument of invention.

There are essays on "experience and geometry" in [Poincaré 1902] and on "probability" in [Poincaré 1902, Poincaré 1908a]. Regarding probability, the need for another type of mathematical convention arises.

Intuitionism in mathematics became important through the ideas of L. E. J. Brouwer. The emphasis on intuition in mathematical thinking does not make Poincaré a precursor of these ideas, but he certainly influenced Brouwer. Discussions about the foundations of mathematics between intuitionists and those favouring an axiomatic treatment became more intense after Poincaré's death. There were also important new elements raised by mathematical logic, showing fundamental problems in the axiomatic approach; see also [Heinzmann 2010], and for an interesting account of the historical discussions, see [Van Dalen 1999].

6.3   Around Mathematics and Mathematicians

"The future of mathematics," a lecture for the fourth International Congress of Mathematicians in Rome (1908), partly reproduced in [Poincaré 1908a], never received the same attention as the 23 problems formulated by Hilbert on a similar occasion in 1900. Difficult explicit problems are perhaps particularly attractive to mathematicians, who are, generally speaking, more down-to-earth than most people think. But Poincaré's lecture appealed to some young mathematicians. L. E. J. Brouwer (1881–1966), who completed his thesis in 1907, attended the congress and wrote to his thesis adviser, D. J. Korteweg [Brouwer 1908]:

> To be able to raise oneself to a view from where one can produce a lecture such as
> Poincaré's lecture "L'Avenir des Mathématiques," whose truthfulness everyone experiences
> and accepts as a guide in his work, this seems to me the highest ideal for any mathematician.

Poincaré's article does not formulate many problems but is instead more philosophical. One of his statements is that much mathematics remains to be developed, something most mathematicians are fully aware of, but it is a fact that is surprising to many, since to one not at the forefront of research, mathematics can appear to be a finished body of knowledge. However, he adds that there has to be a direction in research, a sense of economy that will enable us to make greater steps forward. One of the pitfalls can be the requirement of exactness, producing exceedingly long and dreary papers; also here, economy is required without becoming less precise. A good example of this point of view is the introduction of groups and isomorphisms, which helped enormously to identify similar but very different-looking parts of mathematical theory.

Another point is our attitude with respect to questions from physics and engineering. When a mathematician is asked to solve a certain equation, he will probably answer that there are only a few equations that can be solved, and that this is not one of them. What the mathematician should do, however, is outline qualitatively the behaviour and characteristics of the solution and then take recourse to approximation methods. In Poincaré's time these were series expansions, to which we now add numerical techniques. In fact, today's applied mathematicians are following Poincaré's advice.

In conclusion, a number of more concrete problems are discussed in algebra, topology (analysis situs), and the foundations of mathematics. The discussion of the directions research can take were also addressed in [Poincaré 1908a], in the essay "The selection of facts." Both in mathematics and in the natural sciences, there is an unbounded number of facts. How do we select facts to develop science in a fruitful way? There is a hierarchy of facts, and of interest are those that can be noticed and used more than once. They should not be incidental.

In [Poincaré 1913], we find a discussion of topology, "analysis situs," in the chapter *Pourquoi l'espace a trois dimensions* (Why space has three dimensions). Apart from metric geometry, which deals with distances and coordinates, and projective geometry, which deals with possible transformations and mappings of geometrical objects, we have as a third possibility analysis situs, or "rubber geometry"; we return to this topic in Chapter 10 in greater detail.

Interestingly, Poincaré considers analysis situs a subject in which the role of intuition is very clear. A reason is that the experience of three-dimensional space guides our basic definitions and propositions, while this experience becomes even more important when we are considering the geometry of more than three dimensions.

This part of geometry is concerned with the equivalence of figures when continuous deformation is permitted. So a circle is equivalent to an ellipse, in fact to any closed curve, since we can deform such a curve continuously into a circle; a sphere in three-dimensional space is equivalent to an ellipsoid and to a cube, and so on. But a sphere is not equivalent to a torus, for the hole in the torus prevents continuous deformation into a sphere.

Today, topological notions such as dimension would not be considered philosophical. In Poincaré's time, however, such was definitely the case. Before the treatment of dimension, for example, became part of topology, philosophers and scientists discussed it in a mixture of mathematical, physical, and philosophical terms.

6.4 The Principles of Natural Science

The natural sciences have two basic elements: *experience*, which comes from experiment and observation, and *generalization*. Before describing how we find truth based on these basic elements, we discuss two examples.

Consider a simple experiment: we take a small iron ball and let it fall to the ground. We can repeat the experiment and record falling times, weight of the ball, and other experimental parameters. We will not get exactly the same results from similar experiments, but they will be very close, and we can propose a hypothesis about the gravitational force that is the cause of the ball falling to the ground. The small experimental deviations do not seem to bother us, nor the fact that we can conduct only a finite number of experiments as a means of predicting similar events for an infinite variety of falling objects for all future time. Moreover, we can perform other related experiments with swinging pendulums, for example, and expect that the same hypotheses about the gravitational force will help us to explain and predict the pendulum's motion. What we have described is the interaction of experiment and generalization.

We now consider a much more complicated phenomenon for which we begin not with an experiment, but with a hypothesis: the Earth revolves about the Sun and rotates on its axis. More precisely, we assert that the Earth travels along a nearly circular orbit around the Sun, making one revolution per year while rotating each day about its axis. What we have described is a complex motion, and ours is a hypothesis with which the first-century mathematician and astronomer Ptolemy would disagree. His view was that the Earth stands still at the centre of the cosmos with all the celestial bodies moving about it in circular orbits. There are good arguments for the modern point of view: (1) It is supported not only by observations that verify the position of the Earth and celestial objects as predicted by our hypothesis, but also by the laws of dynamics that we have postulated. (2) The flattening of the Earth at the poles can be explained as a result of axial rotation. (3) The Foucault pendulum behaves in a way that supports the hypothesis of rotation. (4) the large-scale wind circulation on the Earth can be explained by its rotation. (5) Vortical motion, such as that in cyclones and tornadoes, exhibits an orientation that supports the Earth's rotation. And there are many more arguments. In the context of modern science, rotation of the Earth is a *convenient hypothesis*, in contrast to Ptolemy's point of view, which may have been convenient two millennia ago but no longer passes scientific muster.

Note that the term "convenient" differs from the "convention" in mathematics. Unfortunately, Poincaré confusingly uses the term "convention" both for mathematics and for the natural sciences. The essential difference is that hypotheses in the natural sciences are not arbitrary; they are based on experience. We consider this topic in more detail, following [Poincaré 1902] and [Poincaré 1905b].

In considering the natural sciences, one realizes that experience, consisting of experiment and observation, is far from perfect, since we cannot isolate phenomena. For instance, the experimentalist plays a part in perturbing the phenomena under observation, and there are always neglected forces such as, in the case of the falling iron ball, a possible wind force and the friction between the ball and the air. But we may convince ourselves that such factors cause only small errors and that under analogous circumstances there will be analogous outcomes. Secondly, we perform a kind of interpolation: we can perform only a finite number of experiments, but our conclusions are about a continuum of data. For instance, in the case of the falling ball, we expect our results to apply to a continuum of weights, ball sizes, and heights from which the ball is released. With such interpolation we find ourselves already in the second stage, that of theorizing.

Theorizing about experience involves a second ingredient, generalization. This, in turn, leads to the prediction of other outcomes, which we again check by experience. In generalizing, we operate on the fundamental assumption that there is a relatively simple law that explains the experience. For instance, we assume that there exists a gravitational force between masses that is the same in the whole universe and that its operation can be formalized in a relatively simple mathematical equation. This assumption is quite remarkable, considering the complexity of the motions of bodies and the number of particles in the solar system. Therefore, what we assume is basically that we can explain the complexity of experience in nature, the multitude of facts, by a limited number of relatively simple laws.

Science is never complete, and scientific problems are never completely solved. A problem is always solved only to a certain extent and to a certain approximation. Yet the term "convenient hypothesis" that we are using for developing a theory should not be confused with arbitrariness. Notwithstanding the dramatic changes and reformulations of scientific theories and hypotheses, the concepts and laws of science correspond to real relationships because of their emergence from experience. The laws of nature as formulated by science are approximations that improve in quality over time as each generation begins on the foundation laid by the generations that have gone before. As Poincaré would have put it, the laws become more probable. For instance, the concept of energy was attributed first to motion in classical mechanics. Then it was discovered that kinetic energy could be turned into heat, and later that electrical and magnetic fields contain energy. Thus the concept of energy evolved, and the resulting laws of physics became more general and more probable.

We stated that the rotation of the Earth is a convenient hypothesis. Does that mean that one is still free to assume the system propounded by Ptolemy in which the Earth is the immobile centre of the universe? Of course not. In the context of modern science, Ptolemy's hypothesis has probability zero. The hypothesis of rotation has the same certainty as the existence of the observable objects around us. For modern science, Ptolemy's hypothesis is no longer convenient.

Natural science, in contrast to mathematics, is first of all a way of representing experience and of establishing relationships among facts of physical reality. Only in these relationships is there objectivity in science. According to Poincaré, science does not show us the nature of things, the nature of the objects we handle and observe, but the relations between objects.

Consider a mature part of physics, classical mechanics [Poincaré 1902]. It is of interest that there is a difference in style between British and continental physicists. The former consider mechanics an experimental science, while the physicists in continental Europe formulate classical mechanics as a deductive science based on a priori hypotheses. Poincaré agreed with the English viewpoint, and observed that in the continental books on mechanics, it is never made clear which notions are provided by experience, which by mathematical reasoning, which by conventions and hypotheses. Problems that in continental books are either ignored or receive at best an obscure treatment include the following (see also [Poincaré 1905b]):

1. There is no absolute space and corresponding absolute frame of reference.
2. There is no absolute time.
3. We cannot measure equality of time intervals and equality of time at different places.
4. Euclidean geometry is used without discussion, but this geometry is not essential for classical mechanics; it is simply a convenient hypothesis.

Consider as an example the law of inertia: a body on which no force acts moves in a straight line with constant velocity. Is this clear a priori? Certainly not. The law was in fact formulated in fairly modern times. The notion of straight line is based on the convenient hypothesis of Euclidean geometry, while the concept of force or absence thereof is not based on experience but follows from a suitable definition.

The Objective Value of Science

Poincaré's formulation of conventions as convenient hypotheses in the natural sciences produced reactions among philosophers in his time. Poincaré elaborates on these topics in two chapters of [Poincaré 1905b].

Reacting to the mathematician and philosopher Édouard Le Roy (1870–1954), Poincaré asks in [Poincaré 1905b, Chapter 10] whether science is artificial. Le Roy was a student of Henri Bergson and was considered a nominalist. In nominalism, science is considered to be pure convention. This holds both for theory and for facts. The scientist only orders experience; he uses a purely formal language of description (the term nominalism derives from giving names). In this view, all belief in the absolute truth of principles is completely shattered.

Poincaré notes that in this philosophy, not only are facts and theory artificial constructions of the scientist, but the whole scientific enterprise is highly anti-intellectual. In the nominalist view, human intelligence and human language deform everything they touch. Reality is volatile and escapes us the moment we direct our attention to it. Its final consequence is scepticism with regard to all intellectual achievement. For Le Roy, science consists of a system of rules or laws without adding to real knowledge. However, as Poincaré observes, these rules are not arbitrary. They explain experience and they lead to prediction, something that is missing in arbitrary rules. Science is imperfect but involves knowledge that can be used.

Another strange aspect of Le Roy's philosophy is his notion that the scientist creates fact, or more specifically, that there exist a bare fact and also a scientific fact that is created. For instance, we may compare the bare fact that it has become dark in the middle of the day and the scientific fact that an eclipse is occurring. According to Poincaré, the distinction between a bare fact and a scientific one is not always so clear, since some facts can be verified only indirectly. At some point, the difference is merely a matter of the language expressing it. The scientific fact is nothing other than a bare fact translated into another language, and the only thing the scientist creates is the language to describe a fact. In practice, it is even more complicated, for the scientist usually combines several bare facts in order to represent and translate them into a scientific fact.

Poincaré adds that in formulating laws, the scientist has still more freedom. From observing certain regularities, scientists have obtained laws, and then, by unconscious nominalism, principles such as that Newton's law of gravitation reigns everywhere in the universe. Laws can be revised permanently, but for principles we have to adjust or simplify the facts. If all laws have become principles, no science is left; this is a basic limitation of nominalism. Our physical laws change with the conventions that have been chosen. A remaining question is then whether among all these variations of laws there exist universal invariants that are valid independently of the conventions. If such universals can be found, they will exist between bare facts and not between scientific facts, since relations between scientific facts are always governed by conventions.

6.5   Notes on Mathematical Physics

Poincaré's era was no exception to the ever present tendency to view the natural sciences solely for utilitarian purposes, in particular to value them only to the extent that they produce financial profit. Powerful people wanted to limit scientific research to the quest for practical inventions to be used as tools for industrial exploitation. In [Poincaré 1905b, Chapter 5, p. 138], Poincaré defends the autonomy of scientific research:

> Also, science that is carried out only with an eye to applications is impossible; verified results are fertile only when connected to each other in a chain. If one fixes only on those of which one expects immediate results, the connecting components will be missing, and there will no longer be a chain.

Mathematical physics and the relation between its components, namely mathematics and physics, are discussed extensively in [Poincaré 1905b]. In what follows, we present in three parts Poincaré's views on some of these topics. The relation between analysis and physics was also the subject of Poincaré's plenary lecture at the first International Congress of Mathematicians, in Zurich (1897).

Analysis and Physics

Mathematics has three goals: to provide instruments to study nature, to serve a philosophical and aesthetic purpose, and finally to be developed in close collaboration with physicists. Today, experience is at the beginning of all physical considerations, but to express physical laws one needs language. It is in mathematics that one finds the only suitable language. For instance, in the development of thermodynamics, the word "heat" ("chaleur") was unsatisfactory, since it suggests a fixed quantity, a substance, instead of a dynamical state. To find the right word to describe phenomena is of utmost importance. Experience is not precise, but the laws of physics are. They arise out of generalization, and the question is how one chooses from the many possibilities. It is the mathematical spirit that reduces to pure form; for instance, in mathematics one uses the term "multiplication," but it refers to integers, to quaternions, and to many other elements. Quaternions, by the way, emerged as an abstract mathematical construction, but physicists then found a use for them. This is the kind of service mathematicians provide to physics, but to perform this service, mathematicians have to work without immediate preoccupation of usefulness. The mathematician has to work as an artist.

It is instructive to look at examples. In a first example, we can see that a change of language may evoke new generalizations. In Kepler's laws, one saw the geometric and dynamical movements of the planets in elliptical orbits. In the transition to Newton's laws, one began to use differential equations, which made it possible to describe much more complicated gravitational systems than that of the two-body problem.

A second example concerns Maxwell's equations. When Maxwell began his explorations, the equations of electrodynamics explained quite well the experiments of the time. However, Maxwell recognized that the addition of a small term to the equations would make them more symmetric. Later experiments required this term to explain new phenomena. The physicist Maxwell came to this view because he was accustomed to think in terms of symmetries, in the language of vectors and imaginary (complex) quantities.

A third example shows how mathematical analogies between mathematical phenomena assists us in finding new physical phenomena. The Laplace equation describes Newtonian attraction, motion of fluids, electric potentials, magnetic potentials, propagation of heat, and many other phenomena. The images of one field stimulate thinking about similar phenomena in other fields. The concept of flux in hydrodynamics, for instance, was the inspiration for the analogous phenomenon in electrodynamics.

Conversely, what does mathematics obtain from physics? To start with, nature gives impressions and poses questions that we have to answer. In mathematics, there is an infinite number of combinations of objects, so how do we choose a direction of mathematical research? Physics helps us to choose. It also leads us to questions we had not imagined before. Without physics we might be turning in circles. However much variation there might be in the human imagination, nature contains a multitude more such variations.

A simple example can demonstrate this. A natural mathematical object is a whole number, an integer. The exterior world, however, forces us to invent the continuum, which in turn led to differential calculus. Thus the calculus gives perspectives that are missing in number theory. Fourier theory, on the other hand, was originally invented for analysing the propagation of heat, but it is now also used to represent discontinuous functions. Had this problem not arisen naturally, would mathematicians have been enterprising enough to study series leading to discontinuous functions?

The need for other types of series has also arisen from physics. Moreover, the development of the theory of partial differential equations in its many forms was completely triggered by physical phenomena.

As an additional point, physics does more than just suggest problems. It helps to solve them. In the case of the Laplace equation, for instance, the physical phenomena evoked geometric images, which in turn could be used to solve related problems. The most important discoveries are made by guessing before proving. But it should be noted that there are not two types of rigour, a mathematical and a physical rigour. Nonetheless, in physics it helps sometimes to modify the tools. For instance, when analysing certain functions, one can sometimes restrict oneself to polynomials. There are now so many mathematical results that such a modification is often possible.

The History, Present Crisis, and Future of Mathematical Physics

The notes in [Poincaré 1905b, Chapters 7–9] on the present and future of mathematical physics are mostly of historical interest. In line with the discussion of the laws or principles of physics as conventions (or convenient hypotheses) is the observation of the changing formulation of the laws. Earlier, say until 1600, a physical law was something static, something that expressed the internal harmony of the phenomena of the cosmos. Subsequent crises modified and changed those laws. In this context, Poincaré discusses Newton's laws, the principle of relativity with the Michelson–Morley experiment, the Lorentz contraction and the concept of local time, the principle of Lavoisier (conservation of mass), the principle of Mayer (conservation of energy), and the Carnot principle with a discussion of reversible–irreversible.

According to Poincaré, new experiments will overthrow accepted physical principles, and the development of a new mechanics is expected, for instance a mechanics of the dynamics of gases on the basis of statistics.

Science and Reality

The last two chapters of [Poincaré 1905b] are concerned with the nature and role of science in civilization. All natural laws are approximative, but this does not mean that they are accidental. The formulation of laws in natural science means that if certain conditions are satisfied, certain things will probably happen, and progress in science means that the laws formulated will be more and more useful in making predictions about what will happen under given conditions.

One observes, for instance, that an antecedent A produces a consequent B, and one supposes that an antecedent A' that is near to A will produce a consequent B' that is near to B. The assumption is clearly that B depends continuously on A. In this reasoning, one excludes small influences, although phenomena impinging on experiments cannot be completely isolated. Also, one applies induction (the generalization from the specific to the general, not to be confused with mathematical induction as described above). Since there are many similar circumstances in nature, our experiment is not unique, but it can be repeated to a reasonably good approximation. No sequence of the form $A \to B$ will be exactly the same, but many of them will be approximately alike, and we can try to classify them. This is in fact a choice for determinism.

Objectivity in science is guaranteed by contacts with other human beings. These other beings are observing the same things; we are clearly not walking around in a dream. A condition of objectivity in science is that experience be shared by a number of other beings. Transmission is by discourse: no public discussion, no objectivity.

On the other hand, we do not know whether we experience the same things in the same way. Therefore, individually experienced sensations cannot be transmitted. But what we can transmit is the relations between the sensations, the experiences. Objectivity is restricted to relationships.

One might remark that aesthetic emotion is common to most people, but this does not mean that aesthetic sensations are identical but that one has in this respect harmonious and fruitful interactions with other people.

So science is first a classification, a way of approaching and ordering facts. It is a system of relations, and these relations between facts (or objects) can be considered objective. One can ask then two questions:

1. Does science tell us the real nature of things?
2. Does science tell us the real nature of the relations between things?

About the first question, Poincaré wrote [Poincaré 1905b, Section 11.6]:

> About the first question nobody hesitates to answer "no." But I believe one can go further: not only can science not tell us about the nature of things, nothing is able to tell us that, and if some god knew it, he could not find the words to express it. Not only could we not guess the answer, but if one would give it to us, we would not be able to understand it. I even ask myself whether we have a good understanding of the question.

The answer to the second question depends on whether our statements and agreements will persist after us. We can maintain that although theories come and go, there are certain relations that survive all changes. It is somewhat ironic that Poincaré mentions as an example of a persistent notion the ether in the theory of electromagnetic wave transmission. Modern physics has no need for the concept of ether. A better example is the rotation of the Earth as a hypothesis that explains many phenomena. It has the same degree of existence as exterior objects, such as a chair or a table. Kinematically, the statements "the Earth rotates" and "the Earth does not rotate" are both correct. Even stronger, stating that one of the statements is correct and the other incorrect means that one assumes the existence of absolute space and coordinates. But as we have seen before, in physics the statement that the Earth rotates has a much richer content, for it explains many phenomena about which the hypothesis of no rotation tells us nothing. The rotation of the Earth is a convenient hypothesis, as is, by similar reasoning, the motion of the Earth around the Sun.

Finally, one has to perform science because of science. To know all facts is impossible, so we have to choose. Scientists do not choose facts with an eye to useful applications; they select facts that are more interesting or beautiful than others.

Chapter 7
At the End, What Kind of a Man?

Henri Poincaré died on July 17, 1912. During his last year he had been very active, travelling to international conferences and fulfilling many other obligations. On June 26, a few weeks before he died, he gave the public lecture on moral education presented in Chapter 12. The prostate problems that he had experienced at the 1908 international conference in Rome became more serious, and he was advised to have an operation. On Saturday, July 6, he was at a faculty meeting discussing group theory; after that meeting, he told his friend Paul Appell, "Tomorrow I will enter the hospital." The operation took place on July 9 and seemed to have been successful; family members and friends rejoiced and were reassured. A week later an embolism suddenly terminated his life.

The unexpected death of Poincaré was felt as a great loss in France and in the whole scientific world. Paul Appell wrote in [Appell 1925a]:

> The life of Poincaré was an intense and uninterrupted meditation. It was exclusively devoted to scientific work and the family. It will remain a subject of admiration and an example for the youth of France.

In [Darboux 1913], Appell is quoted after Poincaré's death thus:

> Regarding questions, he had the gift of genius to see immediately, including special details, the general idea whence it came and its place in the whole. He also had that unpretentiousness, that distaste for effect, that common sense found in Lorraine, that special good nature that he kept all his life.

Poincaré abhorred prejudice, for instance the assumption that what came from a certain political party was always good and justified and what came from another party always wrong. He did not want to choose between simplified alternatives. In looking for truth there was for him no scepticism and also no given revelation. Religious truth varies across the globe, but science is not diversified; it is a unity. Religion is something for the individual conscience, while scientific truth is the same for everybody and even for all sentient beings in the cosmos. On the other hand, science progresses by small successive approximations. It will never attain final truth for anything.

F. Verhulst, *Henri Poincaré: Impatient Genius*, DOI 10.1007/978-1-4614-2407-9_7,
© Springer Science+Business Media New York 2012

Fig. 7.1 Poincaré at his desk at home

Appell also wrote about Poincaré's patriotism. This may seem outdated to us, but one has to realize that as a boy, Henri witnessed the hostilities of war in 1870–1871, including the loss of Alsace and part of Lorraine to Germany, a loss that was still the reality of those days; those regions were returned to France only in 1918, after the First World War. In the spring of 1912, Poincaré wrote the essay [Poincaré 1913] on ethics and science:

> When one asks us to justify the reasons for our patriotic love, we could become very embarrassed, but imagining our defeated army and invaded France, our heart will be lifted, tears will come into our eyes, and we will no longer listen. And if certain people nowadays put about so many sophisms, it is doubtless because they have not enough imagination. They cannot imagine all the bad things, and if misfortune or some punishment from above opens their eyes, their soul will revolt like ours.

Appell was for many years dean of the Faculty of Sciences at the Sorbonne, and he recalled that Poincaré was quite interested in new academic appointments. He did not automatically vote for the oldest and most experienced candidates but for those he thought had the best scientific qualities. As a colleague he was good-humoured and very conscientious, and he asked very little for himself.

At the request of Mittag-Leffler, Pierre Boutroux (1880–1922), the son of Aline Poincaré and Émile Boutroux, and himself a mathematician and historian of science, wrote about the daily life, habits, and character of his uncle; it was reproduced in [Boutroux 1921]. According to Boutroux, the activity of his uncle's thoughts was all he needed when working at home (see Figure 7.1), in the middle of his family, or in the garden of his summer-house. At work, he usually wrote without much hesitation and with only a few corrections now and then. Very soon after a new article was finished, Poincaré lost interest in it and hardly looked at the proofs sent by the printer. But those were the obvious working sessions. Less obvious was that he always worked: while walking between academic buildings, strolling outdoors in the afternoon, attending a meeting of the Institute, even at social occasions in salons, where he sometimes suddenly interrupted a conversation to follow his thoughts.

Regarding interaction with students and colleagues, it was remarkable that in undertaking new research, he did not believe much in discussion and exchange of ideas. Perhaps that went too slowly for him, but also he saw research as a personal struggle, a confrontation without witnesses. Research was for him a very private affair. It explained in a way why he had few students. His contacts with students took place primarily through his many lectures. Among his few research students, we mentioned earlier Louis Bachelier, Dimitrie Pompeiu, and Mihailo Petrović. Around 1900, there was not much personal contact between professors and students, but already at that time this almost total lack of communication was unusual. Boutroux recalls how, during a visit to Göttingen, he was struck by the lively atmosphere and the many formal and informal discussions of the mathematicians there. Nevertheless, Poincaré was very understanding and sympathetic to beginning students. He always lent them a willing ear when they presented themselves. But when it came to results to be discussed, he was very demanding. If there was not a really new insight, his usual comment was "À quoi bon?" (what is it good for?).

He was not an avid reader of scientific books and articles, and if he read them, it was in a special way. He went directly to the results and then reconstructed the reasoning and arguments himself. His way of thinking was very direct, starting with a problem formulation, jumping long chains of deductive steps to reach a significant result and checking the result afterwards. This also made conversation with him sometimes difficult as he jumped abruptly from one topic to a seemingly unrelated one. Such a mode of thought fits in with the many different scientific topics he dealt with. Poincaré was interested in literature and in geography. He liked to travel with the purpose of seeing famous sites, not as a diversion or in fulfilment of some romantic idea. He rarely went to the same place twice; to see and explore new things was a dominant part of his personality.

Henri Poincaré had a wide interest in culture and science. He was far from being a narrow-minded specialist. This range of interests went with a great sense of responsibility. Once during a conversation, someone mentioned a mathematician who had left his field to do something completely different, adding that this individual probably felt equally fulfilled in his new career as he had as a scientist. Poincaré protested. This is interpreted by Boutroux as the opinion of his uncle that as long as there are things to be explored, one has the obligation to go on. Research was for Poincaré a duty. He could be ironic about many things, but not about science. (Perhaps the mathematician in question was Painlevé, who had a chair at the Sorbonne, began to move into politics in 1906, and left science for politics in 1910.)

A curious but potentially interesting study was made by the psychologist Édouard Toulouse. Poincaré submitted to an analysis of his mental organization by Toulouse, who subsequently wrote a book about it [Toulouse 1910]. The analysis was carried out at the Psychological Laboratory in Paris, resulting in a number of not very exciting observations. For instance:

1. He worked during the same times each day for short periods. Mathematical research took four hours a day, two in the morning and two hours from 5 p.m. till 7 p.m.
2. His normal work habit was to solve a problem mentally and then write it down.
3. He was ambidextrous and near-sighted.
4. He could very well memorize and visualize what he read and heard.
5. He was physically clumsy and not artistically gifted.
6. He was always in a hurry and hated going back for corrections.
7. He believed in letting his unconscious work on a problem while he consciously worked on another problem.

Toulouse examined other prominent and creative citizens as well, for instance Zola, Berthelot, Rodin, and Saint-Saëns.

Gaston Darboux read a eulogy [Darboux 1913] for Henri Poincaré, published in volume 2 of the collected works. About the way he worked, he notes that Poincaré addressed the most difficult problems, preferably in their most general form. When he reached an understanding of a problem he did not turn back to polish up the formulation or arguments but just went on to the next problem. Interestingly, Darboux considered Poincaré's most brilliant results his achievements on Fuchsian and Kleinian functions, which tells also something about the focus of attention among mathematicians around 1910. Regarding the prize memoir for King Oscar II, he understandably makes no mention of the last-minute corrections, but he mentions the criticism by Kronecker of the procedure. According to Darboux, the awarding of this prize was the beginning of Poincaré's name being known to the general public. He became truly widely known to the public through his philosophical writings. Those books were bestsellers and were often translated. *Science et l'hypothèse* sold thousands of copies during its first ten years and was translated into more than twenty languages.

According to Darboux, Poincaré certainly knew the value of what he had accomplished in science, but he never asked for anything special for himself. His reaction to a request was usually positive. For instance, when asked to teach about applications of his theories, he readily complied. When Tisserand died in 1896, Poincaré was teaching mathematical physics, but when asked to switch to Tisserand's mathematical astronomy course, he accepted without demur.

Still, his main interest was in "looking for the truth." He was involved in the supervision of geodetic work, wireless communication, and the council of astronomical observatories. He performed all the tasks this involvement entailed, but he was not really interested in administration.

After reading the opinions of Henri Poincaré's contemporaries, what can we say now, a hundred years after his death?

He was blessed with a happy childhood, a happy marriage, and good friends. As was usual in those times, he kept his personal feelings and affairs within the circle of his family and close friends. Even a long, extended correspondence with Mittag-Leffler contains very little on private matters.

Appell called him good-natured, but he certainly showed fighting spirit, as becomes clear in his exchanges with Felix Klein. But it was a fighting spirit without malice and without lasting hard feelings.

Henri Poincaré was driven by curiosity to understand natural phenomena, to find and formulate mathematical ideas, to understand the workings of the human mind, and to investigate many other topics that piqued his interest. His creativity made for an unusually large scientific output, a lasting legacy. How did he become so productive? What made him a genius? Apart from his emotionally sound and balanced spirit, we can note three factors: his high intelligence, an exceptional memory, and the total and permanent obsession with scientific problems. He truly never stopped thinking about science.

Poincaré was conscientious in his drive to find out and explain, but he was also impatient, perhaps because he had such a wide-ranging vision and was in a hurry to go on to the next problem. He mentioned this impatience himself when discussing graphology and his own handwriting; Pierre Boutroux recalled that Poincaré was not very good at correcting proofs of his papers. Regarding his sense of responsibility with respect to science and society, he was very conscientious. Quite naturally, Poincaré was associated by the public with the establishment of France. This was in line with his prominent social position and because of his family connection to the conservative president of the republic, Raymond Poincaré. However, this does not do him justice. Henri Poincaré spoke out about the dangers of prejudice, for instance in religion and philosophy, and he took a very independent stand in the Dreyfus affair.

Most remarkable about Henri Poincaré are his versatility and his creation of whole new research fields, including the following:

1. Automorphic functions, uniformization
2. The qualitative theory of differential equations
3. Bifurcation theory
4. Asymptotic expansions, normal forms
5. Dynamical systems, integrability
6. Mathematical physics
7. Topology (analysis situs)

In many subjects, Poincaré was in advance of his colleagues, and such individuals are sometimes ignored or even treated with hostility. However, his advanced position did not keep him from the admiration of colleagues and contact with many different people. It might have been otherwise, but he had a good feeling for addressing the great questions of his time and the problems foremost in people's minds. His enormous production helped of course. "Automorphic functions" and "analysis situs" were not likely to appeal to the general public, but "stability of the solar system," "the principle of relativity," and his philosophical essays did.

His style of writing makes his work accessible to scientists from various fields. It contrasts with modern mathematical writing, in particular with some of the pure mathematical schools. But as Darboux and other colleagues at that time noted, there were already objections to Poincaré's style in his own time. As Vito Volterra writes [Volterra et al. 1914], "He is among the scientists as an impressionist among the artists." His writing is like a discourse, he presents an exposition of his ideas about a problem to the reader. The turn of phrase he uses most often in the middle of an article is "ce n'est pas tout" (this is not all). It is true that Poincaré often takes big leaps where nontrivial details have to be filled in; his impatience and his urgent wish to move on show all the time. But the engaging, readable style and the wealth of new ideas more than make up for all that.

Scientists are often in competition, constantly evaluating their colleagues and comparing themselves to them. Richard Feynman wittily called one scientist a "big shot," another one a "small shot." How should Poincaré's stature be evaluated in comparison with scientists of the last 150 years? One name comes immediately to mind: David Hilbert, an eminent mathematician, although more restricted in his choice of research topics. The styles of both men were, however, very different, and trying to compare Hilbert and Poincaré makes little sense. Henri Poincaré was in a class by himself.

Part II
Scientific Details and Documents

Introduction

In this second part we will describe Poincaré's scientific publications in more detail. The emphasis is primarily on differential equations and dynamical systems and secondly on mathematical physics. The chapters on automorphic functions and topology (analysis situs) should be considered as introductory.

A more or less neglected topic here is Poincaré's influence on group theory and algebra. In this respect it is interesting to look at the "Notice sur les travaux scientifiques" [Cartan 1974] of Élie Cartan (1869–1951). Cartan's publications date from 1893 till 1947 and have a wide range. To some extent, his work can be considered an extension and continuation of Poincaré's work on transformation groups and Riemann spaces and the part they play in differential equations and the "new mechanics."

One of the other omitted subjects here is probability. In Poincaré's time, the leading expert in France was Joseph Bertrand. A young mathematician who would become important and productive in this field was Émile Borel. The many books and papers of Borel are still worth reading.

Henri Poincaré was a versatile scientist, and it is impossible to do justice to all his work in one monograph. A number of general references should be useful to the reader. In some cases, these are proceedings of conferences with a number of scientists writing about Poincaré's work; in other cases, they address certain aspects of his work or life. We mention the symposia [Browder 1983, Charpentier et al. 2010], the Solvay workshop of [Novikov 2004], and the accounts [Gray and Walters 1997, Lebon 1912, Volterra et al. 1914, Poincaré 1999].

Chapter 8
Automorphic Functions

The theory of automorphic functions, or Fuchsian functions as Poincaré called them, is a fruitful result of using complex function theory in the analysis of linear ordinary differential equations (ODEs). The early history of its development has been described in [Hadamard 2000].

Elementary knowledge of trigonometric functions gives us periodicity, for instance $\sin(z + 2\pi) = \sin z$, and properties such as the addition rules; this helps in solving a number of linear ordinary differential equations with constant coefficients. What is one to do in the more general cases? Important inspiration came from the study of elliptic functions with the famous example of the solutions of (nonlinear) pendulum equations of the form

$$\ddot{\phi} + f(\phi) = 0,$$

with $f(\phi)$ a cubic polynomial and where we have normalized the physical constants; a dot indicates differentiation with respect to time t, and ϕ is usually the angle of the moving pendulum with the vertical. If the integral of $f(\phi)$ is $F(\phi)$, the energy integral becomes $\frac{1}{2}\dot{\phi}^2 + F(\phi)$, and it is natural to study properties of the inversion of the function defined by

$$t = \int_{\phi_0}^{\phi} \frac{ds}{\sqrt{2E - 2F(s)}}.$$

Here E is a constant of integration, the energy of the oscillator. We would like the integral to define ϕ as a function of t, but generally the inversion causes a number of fundamental problems. Abel and Jacobi discovered the double periodicity of these so-called elliptic functions. Adding a constant ω_1 or ω_2 to the independent variable reproduces the function. Fuchs introduced transformations relating the elliptic functions to linear ODEs with singularities.

F. Verhulst, *Henri Poincaré: Impatient Genius*, DOI 10.1007/978-1-4614-2407-9_8,
© Springer Science+Business Media New York 2012

8.1   From Differential Equations to Automorphic Functions

The inversion problem that arises in solving differential equations is usually illustrated by the following example. Consider the complex function $w = z^2$ (imagine it as the implicit solution of a differential equation for a function of w); inversion looks simple: $z = \sqrt{w}$, or using polar coordinates,

$$w = re^{i\phi}, \quad z = \sqrt{r}e^{i\phi/2}.$$

Start on the real axis at $r = 1$, so $\phi = 0$, $z = w = 1$. Move on a circle around the origin from $\phi = 0$ to $\phi = 2\pi$, producing $z = e^{i\pi} = -1$, a different value. An ingenious solution for the problem of multivaluedness to obtain a unique continuation of such a function was proposed by Riemann. In this example, one notes that an alternative solution of the equation $w = z^2$ is $z = -\sqrt{w}$; this complex function is defined on a second complex plane, called a Riemann sheet. In moving around the origin starting at $\phi = 0$ on the first (complex) Riemann sheet, we join the two Riemann sheets at $\phi = 2\pi$ and continue on the second sheet. The system of two sheets in this example is called a Riemann surface. For algebraic equations in general, there will be a finite number of sheets and a more complicated Riemann surface. An old but still very readable book covering both the elementary and more advanced complex function theory of one variable is [Osgood 1938].

After earlier work of Schwarz and Weierstrass, Fuchs and other mathematicians considered a second-order linear ordinary differential equation of the form

$$y'' + A(z)y' + B(z)y = 0$$

with A and B holomorphic functions of the complex variable z in a region S. There are two independent solutions $y_1(z)$ and $y_2(z)$, and one can consider the ratio $\eta = y_1/y_2$. Fuchs was interested in the behaviour of the solutions near singular points of A and B. He performed analytic continuation of $y_1(z)$ and $y_2(z)$ along a closed curve around such a singularity and inversion of the function $\eta(z)$. This led him to consider a certain linear transformation of η, and more generally to look for functions that are invariant under a substitution of the form

$$z \mapsto \frac{az + b}{cz + d},$$

with complex coefficients. So we have

$$f\left(\frac{az + b}{cz + d}\right) = f(z).$$

The substitutions may act as a discontinuous group. The ratio of the solutions η should be invariant under these linear substitutions, which is a more general property than periodicity. This idea of Fuchs's inspired Poincaré to express the idea at a higher level of abstraction. He called these functions Fuchsian. They are now called automorphic. For his subsequent analysis he had to distinguish between continuous and discontinuous transformation groups. The continuation of these complex functions, the use of Riemann surfaces, and transformations in the complex plane correspond to geometric structures that can be understood only in terms of non-Euclidean geometry. In fact, until Poincaré looked at these problems, non-Euclidean geometry was considered an artificial playground without much relevance to mathematics in general.

To construct Fuchsian groups and the general discontinuous Kleinian groups, Poincaré used series expansions, which led again to new transcendental functions. Altogether, around ten percent of Poincaré's 11 volumes of collected works deals with this topic.

Nowadays, an automorphic function is a meromorphic function of several complex variables that is invariant under a group Γ of analytic transformations of a complex manifold M. Explicitly,

$$f(\gamma(x)) = f(x), \quad x \in M, \quad \gamma \in \Gamma.$$

Automorphic functions of a single complex variable have been extensively studied. The theory also led Poincaré to the formulation of uniformization problems. The integration of algebraic functions and their analytic continuations produce multivalued analytic functions. Uniformization of such functions corresponds to obtaining a parameterization by single-valued meromorphic functions. This topic still contains many fundamental open questions.

The general theory of automorphic functions of Poincaré and the more specific results of Klein and his students produced a new branch of mathematics that is part of complex function theory. It is a beautiful theory, but the question should be raised whether the theory has been important for the development of the theory of ordinary differential equations. The answer is probably negative, but this hardly matters, for the development has led to the relationship between complex function theory and hyperbolic geometry, and also to many results in the study of quadratic forms and arithmetic surfaces.

For Felix Klein, the topic of differential equations remained focused on the use of complex function theory and automorphic functions. Poincaré, on the other hand, first became interested in applications but found it then quite natural to develop powerful generalizations. In this perspective, it is interesting to look at the transcript of a lecture course on differential equations [Klein 1894] given by Klein in 1894. Its additional interest is that it is a copy of a handwritten text of Klein's lectures, probably with a restricted circulation.

8.2   The Lectures on Differential Equations by Felix Klein

The lectures began on April 24, 1894, and ended on August 7, 1894, taking place during the "summer semester" in Göttingen. They were edited by E. Ritter [Klein 1894] and appear in 524 facsimile autograph pages with nice illustrations.

In the preface, Klein notes that the present lectures are a natural sequel to his earlier lectures on hypergeometric functions. He also mentions that in contrast to other authors, he will discuss the global behaviour of solutions, but that this field is so rich that he has to restrict himself to second-order linear ordinary differential equations with three singularities. The emphasis in the discussion is on algebraic and transcendental properties of differential equations, oscillation theorems, and automorphic functions. The treatment is interesting and was already unusual at that time regarding ordinary differential equations, since it is not so much concerned with explicit solutions as with problems of complex function theory such as the role of singularities, Riemann surfaces, and questions of uniformization. Like other treatises on differential equations of that time, it has as its starting point special functions defined by linear second-order ordinary differential equations.

The introduction begins with the equation

$$y'' + py' + qy = 0,$$

with complex coefficients $p(x), q(x)$. The coefficients are algebraic functions on a Riemann plane with regular singularities only. So in a neighbourhood of a singularity $x = a$ that is not a branch point, we can write

$$y(x) = (x - a)^\alpha K(x - a),$$

with $K(x - a)$ a power series in $(x - a)$ and α the "exponent" (nowadays called the index). If $y_1(x)$ and $y_2(x)$ are independent solutions, it is useful to introduce

$$\eta = \frac{y_1}{y_2}$$

and derive an equation for $\eta(x)$; this expression plays a crucial role in the theory to be developed. The properties of $\eta(x)$, think of the doubly periodic elliptic functions, necessitate uniformization, i.e., a transformation of $\eta(x)$ that makes possible the formulation of a suitable function of x and its inverse. The tools for this analysis are provided by the theory of algebraic and transcendental groups.

Klein's philosophy is formulated on pp. 140–146:

> The geometry of the classics is, like all their mathematics, basically synthetic, where one has to understand this word in its old, original meaning. What I mean by this is that from isolated notions a theorem is gradually constructed, and from isolated theorems with difficulty a building up of knowledge; a general theorem is obtained in this way so that subsequently all special cases have been dealt with. (What in modern mathematics is understood by "synthetic geometry" has nothing in common with the old meaning of the word "synthetic";

the modern indication of "synthetic geometry" is only expressing the opposite of "analytic geometry" and means that the synthetic geometry uses its own algorithm, which starts with the consideration of projective sequences of points, whereas the "analytic geometry" uses the algorithms of analysis and algebra. Both are not "synthetic" in a narrow sense.)

After a detailed discussion of analytic and synthetic research, Klein states:

Today, one again uses everywhere in mathematics the synthetic method along with the algorithmic method, and one can distinguish the problems in the separate disciplines by their treatment according to one or the other. I believe one can weigh the value of both methods against each other: with the algorithmic method, if it can be applied at all, one obtains certainly something, even general comprehensive theorems. This is then not so much the merit of the individual mathematician, for he works with the capital of his predecessors, with the supply of ideas which earlier mathematicians have assembled by the creation of the algorithm. It is different with the synthetic method; there everything comes down to having the correct, new thought. There, one does not know whether one will find something, there one has to create one's own path. What one achieves is maybe little, but to a large extent it is the property of the researcher. The algorithm gives progress in an objective respect but not subjectively. One is not so much forced to think independently. The algorithm looks like travelling in a train that goes fast and far, but through cultivated landscapes only; the synthetic method is of the settler who with his axe and much trouble penetrates into the jungle and conquers new domains of culture. In any case the second activity must precede the first.

Klein concludes that in his lectures he will use both algorithmic and synthetic approaches. Algorithmic will be the treatment of algebraic integrability, including algebraic and transcendental groups, and the theory of Lamé polynomials. Synthetic is the discussion of the oscillation theorem (Sturm–Liouville theory) and the theory of automorphic functions, this last chapter taking nearly one hundred pages. It is concerned with the properties of the analytic continuation of the earlier defined function $\eta(x)$, the relation with the geometry of Riemann surfaces, and uniformization questions.

Chapter 9
Differential Equations and Dynamical Systems

9.1 Poincaré's Thesis of 1879

Henri Poincaré presented his thesis to the Faculté des Sciences of the University of Paris to obtain the degree of doctor of mathematical sciences. The title: "Sur les propriétés des fonctions définies par les équations aux différences partielles." It was accepted on August 1, 1879, by a committee consisting of J.-C. Bouquet (chairman), P.-O. Bonnet, and G. Darboux. The text is reproduced in [Poincaré 1916, Vol. 1].

The thesis should be seen as part of Poincaré's work on series expansions of differential equations with singularities while simultaneously putting this in the framework of analysing first-order partial differential equations (PDEs). It is a highly technical and ingenious piece of work, and it is not surprising that the thesis committee needed a relatively long time to reach a decision. The committee asked for examples to illustrate the results, but there are only a few. We will discuss them briefly below. Not only does Poincaré's thesis show his unusual mathematical skill. It also contains a number of new and important concepts. We mention the ideas of an algebroid function and of what is now called a Poincaré domain. A set of eigenvalues, or any set of complex numbers, is considered to be in a Poincaré domain if its convex hull does not contain the origin. This concept also plays a prominent role in what is now called Poincaré–Dulac normalization.

First, we discuss briefly a paper that appeared shortly before the thesis presentation. After the thesis, we discuss the memoirs that appeared very soon thereafter; they can be considered the jump start of the modern theory of ordinary differential equations (ODEs).

F. Verhulst, *Henri Poincaré: Impatient Genius*, DOI 10.1007/978-1-4614-2407-9_9,
© Springer Science+Business Media New York 2012

A Preliminary Study

In [Poincaré 1878], an ODE with singularity in O is considered:

$$x\frac{dy}{dx} = f(x, y),$$

with $f(0, 0) = 0$ and $\partial f / \partial y(0, 0) = \lambda, \lambda \neq 0$, where $f(x, y)$ is holomorphic in x and y in a neighbourhood of O.

Question: what kind of series expansion do we expect for the solutions in a neighbourhood of O? Briot and Bouquet [Briot and Bouquet 1856] proved the existence of holomorphic expansions, the form of which depends on λ (for instance, integer, rational and positive but not integer, real part of λ negative). Poincaré considers cases that are omitted in [Briot and Bouquet 1856], leading to nonholomorphic expansions, i.e., expansion of the solutions in integer powers of x and x^λ:

1. if λ is not a positive integer and has real part positive;
2. if λ is a positive integer.

It is demonstrated that the domain of convergence of such a double series in x and x^λ in the first case is bounded by a circle and a logarithmic spiral. Also, the analysis can easily be extended to higher-order equations, producing, for instance, at second order a triple series.

The Thesis

This first major work that Henri Poincaré produced deals with first-order partial differential equations of the form

$$F\left(z, x_1, \ldots, x_n, \frac{\partial z}{\partial x_1}, \ldots, \frac{\partial z}{\partial x_n}\right) = 0. \tag{9.1}$$

The thesis discusses critical points and singularities of differential equations and is directly tied in with Poincaré's extensive analysis of such problems in his later memoir [Poincaré 1881] and his work on Fuchsian functions. Earlier in the century, Monge, Cauchy, and Jacobi showed that (9.1) could (in principle) be integrated by the method of characteristics, which reduces the analysis of the PDE to the integration of ODEs, the so-called characteristic equations. Since the resulting system of ODEs is generally nonlinear and also may contain singularities, this reduction may still present difficulties. To obtain existence and some quantitative results for ODEs, Briot and Bouquet [Briot and Bouquet 1856] introduced series expansions near a point $z = \beta$ and a point in x-space. It is no restriction of generality to choose $\beta = 0$ and to take the origin O of (x, z)-space ($\mathbb{R}^{n+1}$ or $\mathbb{R}^{n+p}$, depending on the problem) for these points. If $n = 2$, the characteristic equations can be written in the form

$$\frac{dy}{dx} = f(x, y) \quad \text{or} \quad x\frac{dy}{dx} = f(x, y) \quad \text{or} \quad x^m\frac{dy}{dx} = f(x, y)$$

with $f(x, y)$ holomorphic in a neighbourhood of $(0, 0)$ and integer $m > 1$, x and y playing the part of x_1 and x_2.

In the first case, we call $(0, 0)$ an ordinary point of the equation. The second case is nowadays called "weakly singular," while the third case is still largely unexplored even today. Referring to (9.1), the formulation of [Briot and Bouquet 1856] can be stated thus: If p functions $z_1, \ldots, z_p$ of n variables $x_1, \ldots, x_n$ are defined by p equations with terms that are holomorphic near $z = 0$ and $x = 0$ and the functional determinant of the p equations with respect to z does not vanish, then z is holomorphic in x in a neighbourhood of $z = 0$, $x = 0$. So, if we cannot solve the characteristic equations, we can still give a local approximation by series expansion.

The thesis consists of two parts:

1. Problems with the way the solutions of the characteristic equations of (9.1) are defined with *not* simultaneously

$$\frac{\partial F}{\partial p_i} = 0, \quad \frac{\partial F}{\partial x_i} + p_i\frac{\partial F}{\partial z} = 0, \quad i = 1, \ldots, n, \tag{9.2}$$

where we have abbreviated $p_i = \partial z/\partial x_i$.
2. Problems arising from the form of the equations if (9.2) holds.

For his extension of the theory, Poincaré introduces the concept of *algebroid function*: The function z of n variables $x_1, \ldots, x_n$ is algebroid of degree m near $x = 0$ if it satisfies an equation of the form

$$z^m + A_{m-1}z^{m-1} + \cdots + A_1z + A_0 = 0,$$

where the functions $A_0, \ldots, A_{m-1}$ have convergent power series in $x_1, \ldots, x_n$ in a neighbourhood of $x = 0$.

Part 1

In anticipation of solving characteristic equations of the PDE (9.1), we will have a parameterization in t of the variables $x_1, \ldots, x_n$ by the functions $\phi(t), \ldots, \phi_n(t)$. A typical result is then the following.

Lemma 1: If the function z of n variables $x_1, \ldots, x_n$ is algebroid of degree m near $x = 0$ and we introduce the parameterization as above, then z can be expanded in a convergent series in powers of $t^{1/p}$ with natural number $p \leq m$.

Other lemmas explore the various cases that arise for algebroid functions. This is used to study first the construction of solutions of (9.1) in terms of holomorphic functions. In addition, the important problem is considered in which the solution reduces to a given holomorphic function $\beta(x_1, \ldots, x_n)$ if we have $\theta(x_1, \ldots, x_n) = 0$ with θ a given holomorphic function. This opens the way to handling initial–boundary value problems, in both the linear and nonlinear cases.

The cases are illustrated by simple examples. Note that expressions of the form $I(x, y, z)$ defining $z(x, y)$ implicitly are called integrals of the ODEs. For instance, putting $x_1 = x$, $x_2 = y$, consider the following problems:

Example 1

$$\frac{\partial z}{\partial x} + \frac{\partial z}{\partial y} = 1,$$

with the requirement that the integral of the equation reduce to $z = x + x^3$ if $y = x + x^2$. The characteristic equations are easy to solve, and with the boundary condition, we obtain

$$(z - x)^2 = (y - x)^3.$$

The function $z(x, y)$ is algebroid in x and y.

Example 2

$$\frac{\partial z}{\partial x} + \frac{\partial z}{\partial y}(1 - 2z) = 1,$$

with the requirement that the integral of the equation reduce to $x/2$ if $y = x$. With the boundary condition, the characteristic equations produce the solution defined by

$$\left(z^2 + y - x\right) - (z - x) = 0.$$

The integral is holomorphic in z, x, y. Since the equation is satisfied for any z if $x = y = 0$, the surface defined by the integral passes through the z-axis. We cannot conclude at this stage whether $z(x, y)$ is algebroid in x and y.

Example 3

$$\left(\frac{\partial z}{\partial x}\right)^2 + \frac{\partial z}{\partial y} = x + y,$$

with the requirement that the integral of the equation reduce to $x/2$ if $y = x/2$. After a rather long technical analysis, the integral is constructed implicitly; the surface defined by the integral is transversal to the z-axis. The function $z(x, y)$ will be algebroid in x and y.

The theory is supplemented by consideration of linear homogeneous PDEs, linear inhomogeneous PDEs, and an illustration of the complications arising in nonlinear equations.

Part 2

Suppose that the condition given by (9.2) is satisfied. The first case considered is that (9.1) is of the form

$$\sum_{i=1}^{n}(\lambda_i x_i + \cdots)\frac{\partial z}{\partial x_i} = \lambda_1 z,$$

where the ellipsis stands for holomorphic higher-order terms in x_i. We have the following hypotheses:

1. In the complex plane, the convex hull of the points λ_i does not contain the origin (the spectrum is in the Poincaré domain).
2. The numbers λ_i do not satisfy a resonance relation of the form

$$m_2\lambda_2 + m_3\lambda_3 + \cdots + m_n\lambda_n = \lambda_1$$

with $m_2, \ldots, m_n$ positive integers.

These hypotheses lead to the existence of a holomorphic integral and certain series expansions.

A similar analysis is applied to the equation

$$\sum_{i=1}^{n}(\lambda_i x_i + \cdots)\frac{\partial z}{\partial x_i} = \frac{\partial F}{\partial z}$$

and nonlinear equations of a certain form. An explicit example is the PDE

$$\left(\frac{\partial z}{\partial x_3}\right)^2 = \left(\frac{\partial z}{\partial x_1}\right)^2 + \left(\frac{\partial z}{\partial x_2}\right)^2,$$

with a polynomial boundary condition. The solution is implicitly defined in terms of algebroid functions.

Throughout the analysis, many different cases have to be distinguished.

9.2   A Revolutionary Memoir on Differential Equations, 1881–1882

During his assignment at the university of Caen, Poincaré wrote a memoir [Poincaré 1881] that represented a completely new approach to the study of ODEs. Although the memoir is restricted to autonomous second-order equations, the research programme sketched by Poincaré for ODEs is very general, and this programme still dominates research. He writes at the beginning:

Unfortunately, it is evident that in general, these equations cannot be integrated using known functions, for instance using functions defined by quadrature. So, if we were to restrict ourselves to the cases that we could study with definite or indefinite integrals, the extent of our research would be remarkably diminished and the vast majority of questions that present themselves in applications would remain unsolved.

And a few sentences on:

One has already made a first step [in research on ODEs] by studying the proposed function in the neighbourhood of one of the points of the plane. Today we have to proceed much further; we have to study this function in the whole extent of the plane. In this research our starting point will evidently be what we know already of the function studied in a certain region of the plane. The complete study of a function consists of two parts:

1. Qualitative part (to call it thus), or geometric study of the curve defined by the function;
2. Quantitative part, or numerical calculation of the values of the function.

Consider the equation

$$\frac{dx}{X} = \frac{dy}{Y},\tag{9.3}$$

where X, Y are polynomials in x and y, all real. Using the parameter t, (9.3) describes the phase-plane equation of the two-dimensional system

$$\frac{dx}{dt} = X(x, y), \quad \frac{dy}{dt} = Y(x, y).$$

Curves in the (x, y)-plane corresponding to solutions of the equation are called characteristics. For the analysis of (9.3) we use gnomonic projection; this is a cartographic projection of a plane onto a sphere (in cartography, of course, it is the other way around). The plane is tangent to the sphere, and each point of the plane is projected through the centre of the sphere, producing two points on the spherical surface, one on the northern hemisphere, one on the southern. The equatorial plane separates the two hemispheres. Each straight line in the plane projects onto a great circle. So a tangent to a characteristic in the plane projects onto a great circle that has one point in common with the projection of the characteristic on the sphere. Such a point will be called a *contact*. A point on the great circle in the equatorial plane corresponds to infinity.

The advantage of this projection is that the plane is projected onto a compact set, which makes it much more tractable. The price we pay for this is, of course, that we have to give special attention to the equatorial great circle, which corresponds to the points of the plane at infinity. A bounded set in the plane is projected onto two sets, symmetric with respect to the centre of the sphere and located in the two hemispheres.

If at a point x_0, y_0 we have not simultaneously $X = Y = 0$, then x_0, y_0 is a regular point of (9.3), and we can obtain a power series expansion of the solution near x_0, y_0.

If at a point x_0, y_0 we have simultaneously $X = Y = 0$, then x_0, y_0 is a singular point. Under certain nondegeneracy conditions, Poincaré finds four types of singular point, for which he introduces the names saddle, node, focus, and centre, names that remain in use today. These are called singularities of the first type. In the case of certain degeneracies we have singularities of the second type. Points on the equatorial great circle may correspond to singularities at infinity and can be investigated by simple transformations. For instance, if the point is not on the great circle $x = 0$, we transform

$$x = \frac{1}{z}, \quad y = \frac{u}{z},$$

and consider the transformed equation in z and u. If a point on the great circle $x = 0$ is investigated, we transform

$$x = \frac{u}{z}, \quad y = \frac{1}{z}.$$

The next chapter discusses the distribution and the number of singular points. Assuming that the polynomials X and Y are of the same degree and if m indicates the terms of highest degree, and we do not have $xY_m - yX_m = 0$, then the number of singular points is at least 2 (if the curves described by $X = 0$ and $Y = 0$ do not intersect on the two hemispheres, there must be an intersection on the equatorial circle). In addition, it is shown that a singular point on the equator has to be a node or a saddle. In the plane one cannot spiral to or from infinity.

An important concept to be introduced is that of *index*. Consider a closed curve, a cycle, located on one of the hemispheres. In taking one tour of the cycle in the positive sense, the expression Y/X jumps h times from $-\infty$ to $+\infty$; it jumps k times from $+\infty$ to $-\infty$. We call i defined by

$$i = \frac{h - k}{2}$$

the index of the cycle. It is then relatively easy to see that for cycles consisting of regular points, one has the following:

- A cycle with no singular point in its interior has index 0.
- A cycle with exactly one singular point in its interior has index $+1$ if it is a saddle, and has index -1 if it is a node or a focus.
- If N is the number of nodes within a cycle, F the number of foci, C the number of saddles, then the index of the cycle is $C - N - F$.
- If the number of nodes on the equator is $2N'$ and the number of saddles is $2C'$, then the index of the equator is $N' - C' - 1$.
- The total number of singular points on the sphere is $2 + 4n, n = 0, 1, \ldots$.

A characteristic representing a solution of the ODE may touch a curve or cycle at a point. Such a point is called a *contact*; at a contact, the characteristic and the curve have a common tangent. An algebraic curve or cycle has only a finite number of contacts with a characteristic. Counting the number of contacts and the number of intersections for a given curve provides information about the geometry of the characteristics.

A useful tool is the "théorie des conséquents," what is now called the theory of Poincaré mapping. We start with an algebraic curve parameterized by t, so that $(x, y) = (\phi(t), \psi(t))$ with $\phi(t), \psi(t)$ algebraic functions; the endpoints A and B of the curve are given by $t = \alpha$ and $t = \beta$. Assume that the curve AB has no contacts and so has only intersections with the characteristics. Starting at point M_0 with a semicharacteristic, we may end up again on the curve at point M_1, which is the "conséquent" of M_0. Today we would call M_1 the Poincaré map of M_0 under the phase flow of the ODE. Of course, the semicharacteristic may fail to return to AB, for instance because it will swirl around a focus far away or because it ends up at a node. It is also possible to choose the semicharacteristic that moves in the opposite direction and returns to the curve AB in M'; this point is called the "antécédent" of M_0. If $M_0 = M_1$, the characteristic is a cycle, and Poincaré argues that returning maps correspond to either a cycle or a spiralling characteristic. It is possible to discuss various possibilities with regard to the existence of cycles in which the presence or absence of singular points plays a part.

This analysis has important consequences for the theory of limit cycles. A semicharacteristic will be a cycle, a semispiral not ending at a singular point, or a semicharacteristic going to a singular point. Interior and exterior to a limit cycle there has always to be at least one focus or one node. Of the various possibilities considered, it is natural to select annular domains not containing singular points and bounded by cycles without contact. Such annular domains are often used to prove the existence of one or more limit cycles (Poincaré–Bendixson theory).

Five relatively simple examples illustrate the theory. As Poincaré notes, they are simple because the limit cycles in these examples are algebraic. This is instructive but does not represent the general case. For a more general analysis one has to find which regions are acyclic and which ones cyclic. An acyclic region cannot contain a limit cycle; cyclic ones may or may not contain a limit cycle. Poincaré proposes to use an algebraic function $F(x, y)$ that has "nice" properties and whose level curves (he calls them "système topographique") we can study. Among the level curves we may find cycles with and without a contact. In general, traversing a level curve tells us something about the phase flow. Considering the behaviour of the phase flow in polar coordinates may tell us that a cyclic region is monocyclic, i.e., contains exactly one limit cycle. Remarkably enough, with additional assumptions, a few theorems can be formulated.

We discuss briefly the examples for the more general, nonalgebraic, case.

Example 1

Consider the characteristics described in parametric form by the system

$$\frac{dx}{dt} = x(x + y - 2x - 3) - y,$$

$$\frac{dy}{dt} = y(x + y - 2x - 3) + x.$$

It is easy to see that $(0, 0)$ is the only singular point, a focus. It corresponds to two foci on the sphere. Using polar coordinates ρ, ω and transforming $x = \rho \cos \omega$, $y = \rho \sin \omega$, the phase flow is described by

$$\frac{d\rho}{d\omega} = \rho(\rho - 2\rho \cos \omega - 3).$$

Consider the level curves of the family of circles given by $\rho = $ constant. The curve of contacts of this family is the ellipse

$$\rho - 2\rho \cos \omega - 3 = 0.$$

The origin lies within the ellipse. It is clear that the regions $\rho < 1$ and $\rho > 3$ are acyclic. The region $1 < \rho < 3$ has to be studied. The cycles $\rho = 1$ and $\rho = 3$ have opposite signs for $d\rho/d\omega$, so this region is cyclic, and it can be shown that it is monocyclic. The implication is that there exists exactly one limit cycle in the region $1 < \rho < 3$. The equator corresponds to a limit cycle at infinity in the plane.

Example 2

The system is

$$\frac{dx}{dt} = -y + 2x(x + y - 4x + 3),$$

$$\frac{dy}{dt} = x + 2y(x + y - 4x + 3).$$

The origin is a singular point. In polar coordinates, we have

$$\frac{d\rho}{d\omega} = 2\rho(\rho - 4\rho \cos \omega + 3).$$

As in Example 1, the regions $\rho < 1$ and $\rho > 3$ are acyclic, but the cycles $\rho = 1$ and $\rho = 3$ have the same sign for $d\rho/d\omega$, so it is not clear that the annular region $1 < \rho < 3$ is cyclic. Considering the family of cycles $\phi = $ constant given by

$$\phi(\rho, \omega) = \rho - 3, \quad 5\rho \cos \omega = 2,$$

we can demonstrate by calculating derivatives that $\phi = $ constant is always a cycle without a contact in this annular region. The region is acyclic, and the only limit cycle coincides with the equator.

Example 3

A third example shows again a case in which we have an annular region with the same sign for $d\rho/d\omega$ at the boundaries. Using again a suitable cycle, Poincaré shows that there are two subregions where we have monocyclic behaviour, resulting in the existence of two limit cycles in the annular domain. Corresponding to the equator, we also have a limit cycle at infinity.

9.3 Les Méthodes Nouvelles de la Mécanique Céleste

The three volumes of the *Mécanique Céleste* [Poincaré 1892] together form the first modern textbook on dynamical systems. They summarize and extend Poincaré's results in the period 1892–1899 (among which is to be found the prize essay [Poincaré 1890b] for the birthday of King Oscar II), while opening up a completely new way of looking at dynamical systems as described by differential equations and differentiable maps. As in the prize essay, periodic solutions and the question of nonexistence of first integrals receive considerable attention. The applications are often to problems in celestial mechanics and conservative dynamics. The mathematical setting, however, is very general and goes beyond celestial mechanics. For instance, the existence of certain families of periodic solutions and the occurrence of certain bifurcations, such as Hopf bifurcation and the transcritical bifurcation, "emerge" as universal phenomena in dynamical systems.

The solutions of nontrivial dynamical systems can seldom be obtained in terms of elementary functions. Poincaré's answer to this obstruction is to develop qualitative, often geometrical, methods and quantitative approximation methods with a rigorous foundation that go beyond formal calculations.

The thirty-three chapters are numbered as if in one book. The sections, indicated here by §, are also numbered 1–407 consecutively.

Poincaré refers to results of contemporaries, in particular to the papers of Delaunay, Gyldén, Lindstedt, and Hill. Scientists of those times were assumed to have read the *Mécanique Céleste* of Laplace, so those books form a background of results and problems. A reference for celestial mechanics as it was known before 1890 was the volumes of *Mécanique Céleste* by Felix Tisserand [Tisserand 1889]. These four volumes contain an adequate description of the old methods of celestial mechanics. A few biographical details about the scientists mentioned above are given in Chapter 13.

Poincaré's key results in dynamical systems can be listed as follows:

- Poincaré expansion with respect to a small parameter around a particular solution of a differential equation (Chapter 2).
- The Poincaré–Lindstedt expansion method (Chapter 3) as a continuation method and as a bifurcation method for periodic solutions.
- Characteristic exponents and expansion of exponents in the presence of a small parameter; exponents when first integrals exist (Chapter 4).
- The famous proof that in general, for time-independent Hamiltonian systems, no other first integrals exist besides the energy (Chapter 5).
- The idea of "asymptotic series" as opposed to convergent series (Chapters 7 and 8).
- The divergence of series expansions in celestial mechanics (Chapters 9 and 13).
- The Poincaré domain to characterize resonance in normal forms (Chapter 13 and in his thesis, [Poincaré 1916, Vol. 1]).
- The notion of "asymptotic invariant manifold" (Chapter 25).
- The recurrence theorem (Chapter 26).

- The Poincaré map as a tool for dynamical systems (Chapter 27 and [Poincaré 1881]).
- Homoclinic (doubly asymptotic) and heteroclinic solutions; the image of the corresponding orbital structure.

The terminology used for various types of solutions can be confusing. The following list can be useful:

- There are in the restricted three-body problem three *types of periodic solutions* ("sorte de solutions périodiques") that can be obtained by continuation:

  1. The inclinations are zero and all orbits are in one plane; the eccentricities are very small.
  2. The inclinations are zero and the eccentricities are not small.
  3. The inclinations are not zero.

- For linear ODEs of dimension n, a set of n linearly independent solutions is called the "fondamentales." In a neighbourhood of a T-periodic solution, a linearized (variational) system will have solutions of the following form:

  1. If the solution is of the form $e^{\alpha t}\phi(t)$ with $\phi(t)$ T-periodic, it is called a solution of first characteristic ("solution de première espèce").
  2. A solution of a linear system of second characteristic ("solution de deuxième espèce") is $e^{\alpha t} P(t)\phi(t)$ with $P(t)$ polynomial in t.

- The "principal part of the perturbation function" is the part that when expanded in a Fourier series with respect to the angles, contains more than one angle.
- A singular periodic solution (§257) is a solution for which the known integral invariants are linearly dependent when restricted to the periodic solution.
- A T-periodic solution, for convenience called a periodic solution of the first kind ("première genre"), may have nearby kT-periodic solutions with $k > 1$ a positive integer; these are periodic solutions of the second kind ("deuxième genre").
- Periodic solutions of second characteristic ("deuxième espèce") can arise when $F_0(x)$ has a vanishing Jacobian (Chapter 32).

Volume 1: 1892

In the introduction, Poincaré notes that a central problem in the analysis of dynamical systems and in particular for celestial mechanics is the occurrence of secular terms in series expansions for solutions. Serious improvements in the methods to suppress secular terms were obtained by Delaunay, Gyldén, and Lindstedt. Their series expansions for solutions of differential equations, however, are formal. One has no rigorous estimates of the errors of the approximations. Error estimates are essential, not only for mathematical reasons, but also if one wants to decide whether a deviation of observation from calculation corresponds to a real physical phenomenon. To assess the validity of Newton's theory of gravitation, one needs error estimates.

Fig. 9.1 Carl G. J. Jacobi,
mathematician and
mathematical physicist

Chapter 1: Generalities and the Method of Jacobi

This chapter contains basic definitions of ODEs (mostly autonomous), solutions, and integrals. In the earlier literature, solution and integral were used indiscriminately for the same object; here integral is mostly used in the modern sense of "first integral" or "integral of motion." Following Jacobi (see Figure 9.1), canonical (or Hamiltonian) systems are introduced, also canonical changes of variables, i.e., transformations of the variables that conserve the canonical character of the equations. The notation in the three volumes is to use $F(x, y)$ for the Hamiltonian (or energy) function; the conjugate variables x and y are p-vectors. To obtain new canonical variables h, h', Jacobi introduces a generating function $S(y, h)$ with the properties

$$x_i = \frac{\partial S}{\partial y_i}, \quad h'_i = \frac{\partial S}{\partial h_i}, \quad i = 1, 2, \ldots, p.$$

Examples are the Newtonian gravitational two-body and gravitational n-body problems. In three-dimensional physical space, the latter has $3n$ degrees of freedom. The corresponding system of ODEs has dimension $6n$. So the system of ODEs describing the gravitational three-body problem has nine degrees of freedom and dimension 18. Each independent first integral of the system can be used to reduce the dimension by two. Independence of integrals should be understood in the sense of functional independence and of involution using Poisson brackets. The so-called integral of Jacobi for the three-body problem is the energy function (Hamiltonian) written in coordinates relative to the rotating motion of the masses. Other reductions are possible by additional assumptions, in the case of the three-body problem, for instance, by assuming motion of the bodies in a plane or smallness of some of the masses, the restricted three-body problem, or both.

In §9, a relatively simple model of the three-body problem is formulated that will be used as an example throughout the three volumes. Consider two masses,

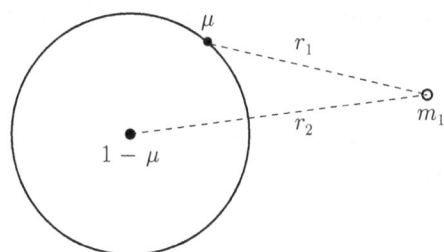

Fig. 9.2 The restricted plane circular three-body problem of §9 (in relative coordinates), which has often been used as a basic example. The mass parameter μ is positive and small, and the mass m_1 is so small that it does not influence the two circular Keplerian orbits of masses $1 - \mu$ and μ

of sizes $1 - \mu$ and μ, moving in concentric circles around their common centre of gravity. Their constant distance is 1; the radii of the circles are normalized to be respectively μ and $1 - \mu$. Suppose now that in the plane of motion of these two masses a third body moves, mass m_1, that is so small that the effect of its motion on the first two masses cannot be observed (for instance, a space vehicle in the Earth–Moon system); see Figure 9.2. Putting the mass $1 - \mu$ in the centre and if the positions of m_1 are (x_1, x_2) and the momenta (y_1, y_2), then the energy function for the motion of the third mass will be

$$\frac{y_1^2}{2m_1} + \frac{y_2^2}{2m_1} - \frac{m_1 \mu}{r_1} - \frac{m_1(1 - \mu)}{r_2}.$$

The numbers r_1, r_2 are the distances from m_1 to the masses μ and $1 - \mu$ respectively:

$$r_1 = (x_2 - (1 - \mu) \sin t)^2 + (x_1 - (1 - \mu) \cos t)^2,$$
$$r_2 = (x_2 + \mu \sin t)^2 + (x_1 + \mu \cos t)^2.$$

Relative coordinates simplify the equations; see Figure 9.2. A canonical transformation to the variables x', y' and dropping the accents leads to the Hamiltonian

$$F(x_1, x_2, y_1, y_2) = F_0(x_1, x_2) + \mu F_1(x_1, x_2, y_1, y_2) + \mu^2 F_2(x_1, x_2, y_1, y_2) \cdots,$$

where

$$F_0 = \frac{1}{2x_1^2} + x_2.$$

A general problem formulation in §14 is to have a Hamiltonian (energy) function $F(x, y)$, where x indicates the positions, y the conjugate momenta. The elements of the n-vector x are x_i, $i = 1, \ldots, n$, etc., and the equations are

$$\frac{dx_i}{dt} = \frac{\partial F}{\partial y_i}, \qquad \frac{dy_i}{dt} = -\frac{\partial F}{\partial x_i},$$

with a convergent expansion in the small parameter μ of the form

$$F(x, y) = F_0(x) + \mu F_1(x, y) + \mu^2 F_2(x, y) + \cdots,$$

where $F_1, F_2, \ldots$ are 2π-periodic in y. The formulation of the general three-body problem is discussed in §§11–19.

Chapter 2: Series Expansions

This is a basic chapter that deals with series expansions of solutions of ODEs. After discussing Cauchy's classical results for series expansions of differential equations, Poincaré formulates his celebrated expansion theorem. The idea is as follows: Consider the n-dimensional system

$$\frac{dx}{dt} = X(x, t, \mu),$$

where X is T-periodic in t and μ is a vector of small parameters. A particular solution of the equation

$$\frac{dx}{dt} = X(x, t, 0)$$

is $\phi(t)$. The vector field X has convergent series expansions with respect to the small parameter(s) μ and $x - \phi(t)$. Then a solution in a neighbourhood of $\phi(t)$ can be expanded in a convergent series of the form

$$x(t) = \sum_{n=0}^{\infty} \mu^n c_n(t).$$

Assuming that the solution does not run into a singularity, the coefficients $c_n(t)$ are uniformly bounded, and so the expansion is uniformly valid in time. The last statement has importance for existence questions but not for quantitative results, since in practice one can calculate a finite number of terms only.

The proof uses majorizing series; see [Verhulst 2000] for a modern proof. This Poincaré expansion theorem should be distinguished from the Poincaré–Lindstedt expansion theorem, which is based on it but aims at obtaining series expansions for periodic solutions.

The remaining part of the chapter is concerned with a summary of Floquet theory, the theory of implicit functions, and Puiseux expansion for the solutions of algebraic equations.

Comments on Chapters 1 and 2

The chapters begin by summarizing the theory of ODEs and Hamiltonian systems as they were known at the end of the nineteenth century. This starting point for the three volumes involves series expansion for ODEs, Hamiltonian equations and canonical transformations, Floquet theory, the theory of implicit functions, and Puiseux expansion for algebraic equations; see [Verhulst 2005, Chapter 15]. A Taylor series of a function with respect to a variable or parameter is a convergent expansion in integral powers. Consider, for example, an algebraic equation in which μ is a small positive parameter:

$$x^2 + \mu x - 1 = 0.$$

If $\mu = 0$, there are two distinct roots. The Taylor expansion of the roots with respect to μ is

$$x = \pm 1 - \frac{1}{2}\mu \pm \frac{1}{8}\mu^2 + \cdots .$$

Puiseux expansion takes place with respect to rational powers of a variable or parameter. Consider the example

$$x^2 + \mu x - \mu = 0,$$

with multiple roots if $\mu = 0$. We have the Puiseux expansion for the roots

$$x = \pm \mu^{1/2} - \frac{1}{2}\mu \pm \frac{1}{8}\mu^{5/2} + \cdots .$$

The occurrence of secular terms in a convergent expansion can be demonstrated by Taylor expansion of the periodic function

$$\sin(1 + \mu)t = \sin t + \mu t \cos t - \frac{1}{2}\mu^2 t^2 \sin t + \cdots .$$

The radius of convergence is positive, but μt has to be small.

Chapter 2 contains the first formulation (1892) of the Poincaré expansion theorem, stating and proving that under certain assumptions, expansion with respect to a small parameter can be achieved in a neighbourhood of a solution of a differential equation. Very soon afterwards, Émile Picard, a colleague of Poincaré in Paris, included this in his treatise on analysis [Picard 1891, Vol. 3, Chapter 8, Section 1]. Picard's proof is somewhat simpler, for he uses his successive approximation scheme "Picard iteration," which is basically a contraction procedure. A more abstract version can be found in [Verhulst 2000]. Picard adds that there is no obstruction to generalizing the result to the case of expansion with respect to more than one small parameter.

Chapter 3: Periodic Solutions

Suppose we have an ODE with a small parameter μ in the form

$$\dot{x} = X_0(x,t) + \mu X_1(x,t) + \mu^2 X_2(x,t) + \cdots,$$

where x is an n-vector, the right-hand side represents a convergent power series expansion with respect to μ, and the X_i are holomorphic with respect to x and t in a certain region; either they do not depend explicitly on t (the autonomous case) or they are 2π-periodic in t. Suppose that the equation contains a periodic solution for $\mu = 0$. Can this periodic solution be continued for small values of μ?

Poincaré reduces this problem to the question of unique solvability of a system of n algebraic or transcendental equations with n unknowns requiring the determinant of the Jacobian of the system to be nonzero. The equations are obtained by expanding the solutions for $\mu > 0$ in a neighbourhood of the periodic solution at $\mu = 0$ according to the expansion theorem of Chapter 2 and then applying the periodicity condition. This allows application of the implicit function theorem and is the basis for what is today called the Poincaré–Lindstedt method; for an introduction, see [Verhulst 2000].

What happens if a solution is multiple of order m? Suppose that for small positive μ, the number of solutions is m_1, and for small negative μ, the number is m_2. The numbers m, m_1, m_2 are expected to have the same parity (all of them are either odd or even). If $m_1 \neq m_2$, the difference is even, so if μ passes through zero, at least two solutions vanish or emerge. A periodic solution can vanish only by merging with another periodic solution.

Suppose (as in §37) that the period is fixed as the (nonautonomous) vector field depends periodically on t. The n-dimensional parameter β will indicate the perturbation of the initial values of the periodic solution at $\mu = 0$. In our system of n algebraic or transcendental equations, we can eliminate $n - 1$ variables to obtain an equation of the form

$$\phi(\beta, \mu) = 0.$$

This relation can be represented by a curve in a plane with each point corresponding to a periodic solution; this is nowadays called a *bifurcation set*. As we have seen, there are various possibilities, depending on what happens if $\mu = 0$. If for $\mu = 0$ the relation is identically satisfied, we have an infinite number of periodic solutions for $\mu = 0$. The implication is that μ is a factor, and we can write $\phi = \mu \phi_1(\beta, \mu)$. In this case, there are at least two bifurcation branches: for $\mu = 0$ and $\phi_1 = 0$. If the branches are not tangent but intersect transversally, the intersection corresponds to a multiple solution.

Other observations can be made without looking at explicit ODEs. An important case in applications is that our original system of ODEs has a first integral—in celestial mechanics, the energy integral. The first integral may be of a very general form $F(x,t) = $ constant. In this case, we will reduce our system to a system of lower dimension, and a periodic solution arising from this system corresponds to a family of periodic solutions of the original system.

There is a modification of the theory if the system of ODEs is autonomous (§38). In this case, the period of the solution is not a priori fixed, and if we find for $\mu = 0$ a T-periodic solution, we will look for a periodic solution for $\mu > 0$ with a period slightly perturbed with respect to T. This adds a small parameter to the equations arising from the periodicity condition. If, in addition, the autonomous system has a first integral $F(x) =$ constant, we expect to find a family of periodic solutions.

How does one apply this to the three-body problem? Assume that two of the masses, m_2 and m_3, are small, so that for $\mu = 0$, these masses describe Keplerian orbits around the large mass m_1. We expect three types of solutions:

1. The inclinations are zero (all orbits are in one plane) and the eccentricities are very small.
2. The inclinations are zero and the eccentricities are not small.
3. The inclinations are not zero.

Solutions of the First Type

In this chapter, most of the attention is given to solutions of the *first type*. For $\mu = 0$, there are two first integrals, energy and angular momentum, and the periodicity condition will be with respect not to time but to the longitude angle. By showing that a certain functional determinant is nonzero and by applying the implicit function theorem, it is demonstrated that there exists an infinite number of periodic solutions of the first type. The analysis simplifies somewhat if one of the small masses is considered to be zero, i.e., its motion does not influence the motion of the other two masses. Such an analysis was made by Hill in considering the Sun–Earth–Moon system. While praising Hill's achievements, Poincaré offers some corrections to his analysis (§41).

More generally, one can consider the Hamiltonian equations of motion with Hamiltonian $F(x, y) = F_0 + \mu F_1 + \mu^2 F_2 + \cdots$, where F is periodic in y, and F_0 depends on x only. If the determinant of the Jacobian $|\partial F_0/\partial x|$ is nonzero, then the existence of periodic solutions is possible, depending, of course, on F_1. In practice, this Jacobian determinant may be singular, as happens, for instance, in the three-body problem, and additional assumptions are necessary. Poincaré considers in §43 the case in which F_0 does not depend on some components of x, producing an explicit series expansion and a proof of convergence by majorization.

Solutions of the Second Type

Consider again solutions of the three-body problem moving in a plane but with nonzero eccentricities of the limiting ($\mu = 0$) Keplerian orbits. We have six position variables and six conjugate momenta. Put again for the Hamiltonian

$$F(x, y) = F_0 + \mu F_1 + \mu^2 F_2 + \cdots,$$

where F_0 depends only on two position variables. The solutions of the system with $\mu = 0$ are chosen with two initial values such that they are 2π-periodic. Looking for periodic solutions if μ is nonzero but small, we expand F_1 in a Fourier series with respect to the three angles. Averaging the Fourier series over the angles and substituting the two known initial values, we retain the periodicity conditions for the remaining four variables. Analysis of this problem for three-body periodic solutions of the second type shows that there are two sets of solutions, corresponding to periodic solutions that are in symmetric conjunction.

Interestingly, Poincaré notes that a problem can arise in considering other perturbations of the Keplerian orbits. This can be remedied by employing a slightly different system of variables.

Solutions of the Third Type

The analysis of solutions of the third type begins in the same way as for periodic solutions of the second type, except that the inclinations are nonzero, so that we have eight variables. Using the Fourier expansion of F_1 and averaging over the angles, we obtain two periodic solutions by continuation of the inclination-zero case. An important question is then whether there exist periodic solutions with nonzero inclination that do not have the inclination-zero solutions as a limit. For the calculation, one needs more terms of the averaged Fourier expansion, in fact terms to the third degree in the inclination i and eccentricity e. Using the expressions provided by Tisserand [Tisserand 1889], Poincaré obtains periodic solutions of type three that do not have solutions of type two as a limit.

Concluding Remarks

1. In practice, a celestial body will not have initial conditions that result in periodic motion. But periodic solutions are the beginning for exploring other types of solutions that have nearby initial conditions (see subsequent chapters).
2. If a celestial body like the Moon describes near-periodic motion, then a suitable periodic solution describes a certain ideal motion with deviations that can be estimated. This idea was used by Hill for the motion of the Moon (§49).
3. In the case of resonance, certain periodic solutions may cease to exist, but close to the resonance value, the same solutions may persist (§50). This happens, for instance, for the satellites Hyperion and Titan of Saturn, which are in near $3/4$ resonance. Another analysis following this idea was carried out by Laplace for three satellites of Jupiter.
4. Very different types of dynamics are possible. Suppose, for instance, that two small bodies revolve around a large body in near-Keplerian orbits E and E'. At some time they approach each other, and their mutual attraction has to be taken into account; this results in two new near-Keplerian orbits E_1 and E_1'. Another

near-collision may take place, resulting in new near-Keplerian orbits or perhaps in the original E and E'. In this way, a new type of periodic solution may arise.

5. A special case is the consideration of periodic solutions near a point of equilibrium that exists for any value of the small parameter (§51). The corresponding stationary solution can be considered periodic with arbitrary period. Looking for a periodic solution with prescribed period T, we can formulate periodicity conditions that can be analysed. In this way, we find *periodic solutions of the second kind*, which will be studied more extensively in a subsequent volume (Chapters 28–31). This emergence of a periodic solution near an equilibrium happens when two eigenvalues are (conjugate) imaginary. The phenomenon is nowadays called Hopf bifurcation; because of its prominence in dynamical systems theory, it gets a separate description in Section 9.4.

6. Finally, one can consider the lunar solution of Laplace: constant in conjunction or in opposition and at constant distance. A nearby periodic solution can be found with an expansion in fractional powers of the small parameter.

Comments on Chapter 3

This chapter on periodic solutions is long (pp. 79–161). It combines for the first time the existence and approximation of periodic solutions by perturbation theory, and it opens up another new field: bifurcation theory for ODEs. The theory is very general and applies both to Hamiltonian and to dissipative systems. It was known from examples in certain differential equations that branching into multiple solutions was possible. For instance, in the celebrated problem of self-gravitating rotating fluid masses, this occurs at certain values of the rotational velocity (see Section 11.2). The treatment in Chapter 3, however, is very general and not concerned with examples. It is based on the implicit function theorem, which yields the continuation method for periodic solutions, usually called the Poincaré–Lindstedt method.

Lindstedt formulated the expansion method for a few explicit equations. His procedure is entirely formal but ingenious. Interestingly, in Lindstedt's calculations, one can recognize already the notion of "multiple scales," but there is no indication that this was his view of the phenomena in question.

Émile Picard, see Figure 9.3, gave a clear account of Poincaré's method in 1896, including the case that a first integral exists, the case that the system is autonomous, and the case that an equilibrium has purely imaginary eigenvalues (Hopf bifurcation) [Picard 1891, Vol. 3, Chapter 8]. In his discussion, Picard adds two interesting observations. First, he notes that one can also study in this way periodic solutions of the system

$$\dot{x} = X(x, y, \mu),$$
$$\dot{y} = Y(x, y, \mu),$$

where x is a p-vector, y a q-vector, and X and Y are 2π-periodic in y. One looks for ω-periodic solutions $x = \phi(t)$, $y = \psi(t)$ with additionally $\psi(t+\omega) = \psi(t)+2k\pi$,

Fig. 9.3 Charles Émile
Picard, mathematician

k a natural number. The period ω is close to the period of a known periodic solution
if $\mu = 0$.

A second observation made by Picard is that in the case of an equilibrium, one
does not need a small parameter μ to obtain periodic solutions. The interpretation
of this observation is probably as follows. Consider the n-dimensional system

$$\dot{x} = X(x), \quad X(0) = 0.$$

Suppose that $X(x)$ can be expanded in a convergent power series with respect to
the components of x and that the equilibrium $x = 0$ has two purely imaginary
eigenvalues. Expansion of $X(x)$ produces, near $x = 0$,

$$\dot{x} = Ax + F(x),$$

where $F(x)$ represents a Taylor expansion beginning with quadratic terms, and A is
a constant $n \times n$ matrix. Considering a neighbourhood of $x = 0$, we rescale $x \mapsto \mu x$
with μ a small parameter. Dividing by μ, we obtain

$$\dot{x} = Ax + \mu F(x, \mu).$$

For this equation we can start our search for a periodic solution in a μ-neighbourhood
of $x = 0$.

Poincaré's calculations of a periodic solution produce a curve in parameter space
that is called a *bifurcation set*. There are modifications for cases of Hamiltonian
and autonomous systems. In applying this to the three-body problem, several new
aspects arise. One can distinguish three types of periodic solutions with in each case
a different continuation analysis. Moreover, in such practical problems there are
cases in which the Jacobian determinant $|\partial F_0/\partial x|$ vanishes, which requires a special
approach. The implicit function theorem does not automatically apply here, but the
analysis for obtaining periodic solutions can still be carried through. In Section §51,

the framework is sketched for what later will be called *Hopf bifurcation*. The analysis of the orbits of the satellites of Jupiter was continued later by Willem de Sitter (1872–1934) [de Sitter 1907].

In a letter of January 1892 [Poincaré 2012], Hill thanks Poincaré for sending him the first volume of his celestial mechanics book; he agrees with the corrections to his papers on lunar motion proposed in the book.

Chapter 4: Characteristic Exponents

Consider an n-dimensional autonomous equation of the form

$$\dot{x} = X(x),$$

and suppose we know a particular solution $x = \phi(t)$. We call this a generating solution. In studying neighbouring solutions of $\phi(t)$, we put

$$x = \phi(t) + \xi.$$

The variational equations of $\phi(t)$ are obtained by substituting $x = \phi(t) + \xi$ into the differential equation and linearizing for small ξ to obtain

$$\dot{\xi} = \frac{\partial X}{\partial x}\bigg|_{x=\phi(t)} \xi.$$

One solution of this linear system is $\dot{\phi}(t)$.

Suppose now in addition that the original equation for x has a first integral

$$F(x) = \text{constant}.$$

For the particular solution, one has $F(\phi(t)) = c_1$, and for the neighbouring solutions,

$$F(\phi(t) + \xi) = c_2.$$

Expanding and linearizing for small ξ leads to the integral of the variational equations

$$\frac{\partial F}{\partial x}\bigg|_{x=\phi(t)} \xi = c_2 - c_1.$$

Hill used such an approach for the motion of the Moon by beginning with a periodic solution $\phi(t)$ of the first type; the results agree with observations.

Variational Equations in the Case of the Hamiltonian Equations of Dynamics

Beginning in §56 with the $2n$ Hamiltonian equations of motion

$$\dot{x} = \frac{\partial F}{\partial y}, \quad \dot{y} = -\frac{\partial F}{\partial x},$$

and a particular solution (x, y), it easy to write down the variational equations with respect to this solution. Suppose now that we have two solutions, (ξ, η) and (ϕ, ψ), of the variational equations. It is easy to show that they have to obey the relation

$$\sum_{i=1}^{n}(\psi_i \xi_i - \phi_i \eta_i) = \text{constant},$$

which can be generalized for $2p$ independent solutions. This can be used to construct integrals of the variational equations. For if $(f(t), g(t))$ is a particular solution and (ξ, η) an arbitrary solution, we have the time-dependent integral of the variational equations

$$\sum_{i=1}^{n}(g_i \xi_i - f_i \eta_i) = \text{constant}.$$

In a similar way, we can construct individual solutions if an integral of motion is given. Moreover, we can relate this to the classical integrals of motion.

Poincaré uses some space to explain linear algebraic transformations, which were not well known at his time, and he defines *characteristic exponents* of a periodic solution. These are obtained by linearizing around a T-periodic solution of an autonomous system; the resulting variational equations are linear equations with T-periodic coefficients. Their solutions are of the form (today called the Floquet decomposition)

$$\xi = e^{\alpha t} S(t)$$

with $S(t)$ T-periodic and α a complex number, called a characteristic exponent of the periodic solution. The real parts of the exponents determine the stability of the periodic solution. With regard to stability, the characteristic exponents of a periodic solution play the same part as the eigenvalues of an equilibrium point. Note that in this section (§59), the term "integral" is used with the old meaning of solution.

Consider again the n-dimensional equation with T-periodic right-hand side

$$\dot{x} = X(x, t)$$

and assume that we know a generating (nonconstant) T-periodic solution $\phi(t)$. Using the variational equations, we can obtain a linear system of equations whose characteristic eigenvalue equation produces the characteristic exponents. There are some important cases:

- It is clear from the linear system determining the characteristic exponents that if $X(x,t)$ does not depend explicitly on t, the autonomous case, then one of the characteristic exponents is zero.
- If the vector field $X(x,t)$ contains a small parameter μ, and if it can be expanded with respect to this parameter and admits a T-periodic solution $\phi(t)$ for $\mu = 0$, then a periodic solution for small nonzero values of μ exists if all the characteristic exponents of $\phi(t)$ are nonzero.
- If in the preceding case the vector field X is autonomous and we have one and only one zero characteristic exponent, then the same conclusion for the existence of a periodic solution holds.
- If we have a T-periodic equation $\dot{x} = X(x,t)$ with T-periodic solution $\phi(t)$ and in addition an analytic first integral $F(x) = $ constant, then at least one of the characteristic exponents of $\phi(t)$ is zero. The rather exceptional case for this result is that all the partial derivatives $\partial F/\partial x$ vanish for $x = \phi(t)$.
- If the vector field X is autonomous and we have p independent first integrals, $p < n$, then we have at least $p + 1$ characteristic exponents equal to zero.

A number of special results hold in the case that our nonlinear system of differential equations is Hamiltonian and autonomous. In §69 it is proved that in this case, the $2n$ characteristic exponents of a periodic solution emerge in pairs $\lambda_i, -\lambda_i$, equal in magnitude and of opposite sign. In addition, the energy integral produces two characteristic exponents zero; if there exist p other independent first integrals, then either we have $2p + 2$ characteristic exponents zero, or the functional determinants of the integrals restricted to the periodic solution vanish. For the proof, Poincaré uses Poisson brackets and the theory of independent solutions of linear systems.

A result is contained in §74 that will be used again later. Starting with the autonomous (Hamiltonian) equations of dynamics that can be expanded with respect to a small parameter μ, we shift again the equations of motion to a neighbourhood of a given T-periodic solution. Two of the exponents α will vanish, the remaining exponents depend on μ, and the characteristic equation is even and of the form

$$G(\alpha, \mu) = 0.$$

We conclude that the exponents α can be expanded with respect to powers of $\sqrt{\mu}$.

Transformation of the Time Variable

Suppose that we have an autonomous system and we transform $t \mapsto \tau$ by

$$\frac{dt}{d\tau} = \Phi(x).$$

For the original system we have a T-periodic solution $\phi(t)$. In the time variable τ, the period is

$$T' = \int_0^T \frac{dT}{\Phi(\phi(t))}.$$

The new exponents will have the former values multiplied by T/T'.

The last sections (§§74–80) of the chapter are concerned with the possibility of *expansion of the characteristic exponents*. Consider the case of the equation of motion in Hamiltonian form with Hamiltonian expanded in a power series with respect to the small parameter μ: $F = F_0 + \mu F_1 + \mu^2 F_2 + \cdots$; here F_0 depends only on the position x. Assume again that as before, we have a T-periodic solution of the equations of motion, expanded in a power series with respect to μ. The characteristic exponents of the periodic solution correspond to the roots of a $2n$-dimensional eigenvalue equation with two roots equal to zero. An interesting case arising in practice is that F_0 does not depend explicitly on all the components of the position x; in this case, the number of zero characteristic exponents increases correspondingly. Note that in the simplified models of the three-body problem considered earlier, we have only two characteristic exponents equal to zero. In general, for the case of three degrees of freedom, we have four nonzero characteristic exponents, and explicit conditions for existence and stability of periodic solutions are written down.

Comments on Chapter 4

The chapter on characteristic exponents contains very basic material that, strangely enough, more or less vanished from the general mathematical literature on ODEs between 1910 and 1970. It continued to be discussed in engineering books, in particular with respect to problems of parametric excitation. Remarkable is the construction of integrals of linear variational equations near a particular solution. Important is also the form that the expansion of the characteristic exponents with respect to a small parameter can take. This plays a part in stability analysis, which is nearly always based on perturbation theory.

If an autonomous system of ODEs has more than one characteristic exponent equal to zero near a periodic solution, this suggests the existence of a first integral (§65; there is a certain nondegeneracy condition). Consider as an example a nonconstant periodic solution of a time-independent Hamiltonian system. There will be two zero characteristic exponents. Computing the eigenvalues in the case of a system with two degrees of freedom, we have two purely imaginary or two real eigenvalues, corresponding respectively to stability and instability. In the case of a system with three degrees of freedom, we have in general four possibilities; see Figure 9.4.

If the time-independent Hamiltonian system has a periodic solution with more than two zero characteristic exponents, this can be caused by the presence of another first integral besides the energy, or it may be an exceptional case. Assuming a Taylor

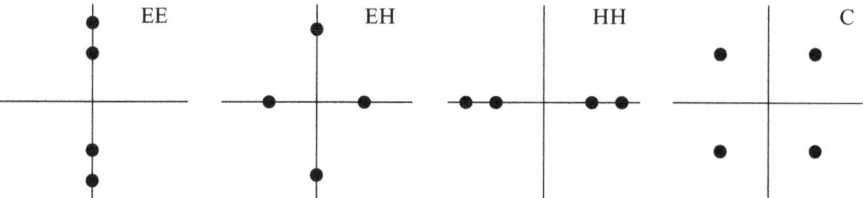

Fig. 9.4 The four possible cases in the complex plane of eigenvalues of a periodic solution of a time-independent Hamiltonian system with three degrees of freedom; the unstable cases are indicated by EH (elliptic–hyperbolic), HH (hyperbolic–hyperbolic), and C (complex); the stable case is EE (elliptic–elliptic), but in that case, one has still to study the possibility that nonlinear terms destroy the stability

expansion near the stable equilibrium, we have for the Hamiltonian near the stable equilibrium

$$H(x, y) = \sum_{i=1}^{n} \frac{1}{2}\omega_i \left(x_i^2 + y_i^2\right) + H_3 + H_4 + \cdots,$$

with H_k homogeneous polynomials of degree k. The first term on the right-hand side is called H_2. In the case of normal forms of Hamiltonian systems near a stable equilibrium, one obtains the quadratic part H_2 of the Hamiltonian as a second first integral (see [Sanders et al. 2007]). But this second integral corresponds to linear terms in the equations of motion, and it represents such an exceptional case. An example of more than two zero characteristic exponents is the normal form of a system with three degrees of freedom in $1 : 2 : 5$ resonance, where normalization to H_3 produces two families of periodic solutions on the energy manifold. The normal form truncated to cubic terms is integrable. The families break up when the normal-form terms from H_4 are added; no low-degree algebraic third integral can be found in this case; see Figure 9.5 and [Van der Aa and De Winkel 1994].

The technical problems connected with drawing conclusions from the presence of more than two zero characteristic exponents have probably prevented its use in research of conservative dynamics, but the statement that "a continuous family of periodic solutions on the energy manifold is a nongeneric phenomenon" is an abiding feature in the literature. Today, the analysis is made easier by the use of numerical continuation methods.

Chapter 5: Nonexistence of Uniform Integrals

A fundamental theorem is formulated and proved in the case of the time-independent $2n$ Hamiltonian equations of motion

$$\dot{x} = \frac{\partial F}{\partial y}, \quad \dot{y} = -\frac{\partial F}{\partial x},$$

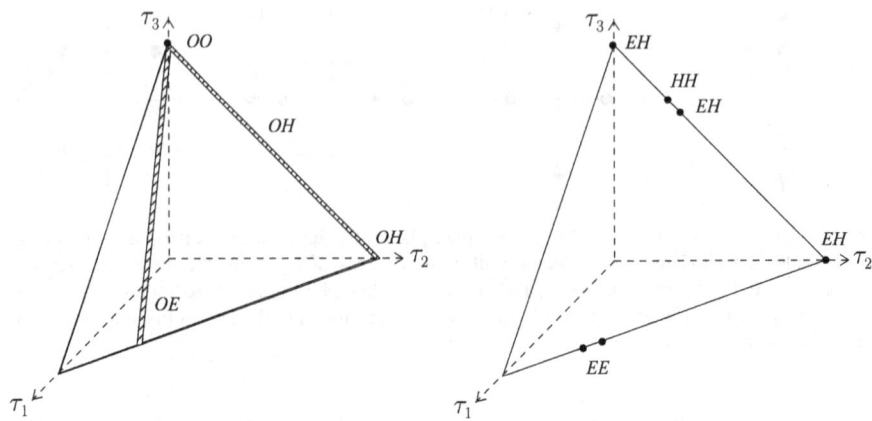

Fig. 9.5 Two action simplices of the Hamiltonian system with three degrees of freedom in $1 : 2 : 5$ resonance. The action simplex shows only the actions of the six variables (actions and angles). The triangle that faces us represents H_2 at fixed energy; a black dot corresponds to a periodic solution. To the left is the simplex of the normal form to H_3 with two continuous families of periodic solutions; the normal form is integrable. To the right is the simplex of the normal form extended to H_4, where the families have broken up into 6 isolated periodic solutions on a given energy level

with small parameter μ and the convergent expansion $F = F_0 + \mu F_1 + \mu^2 F_2 + \cdots$; here F_0 depends on x only and the Jacobian is nonsingular: $|\partial F_0/\partial x| \neq 0$. Suppose $F = F(x, y)$ is analytic and periodic in y in a domain D; $\Phi(x, y)$ is analytic in x, y in D, analytic in μ, and periodic in y:

$$\Phi(x, y) = \Phi_0(x, y) + \mu \Phi_1(x, y) + \mu^2 \Phi_2(x, y) + \cdots .$$

With these assumptions, $\Phi(x, y)$ cannot be an independent first integral of the Hamiltonian equations of motion unless we impose further conditions.

The reasoning contains various steps:

1. If Φ is an integral, we have for the Poisson bracket $[F, \Phi] = 0$, so that by expansion, we have successively $[F_0, \Phi_0] = 0$, $[F_1, \Phi_0] + [F_0, \Phi_1] = 0$, etc. One can prove that Φ_0 is not a function of F_0.
2. Using Fourier expansion of Φ_0 and the Hessian of F_0 being nonzero, it follows that Φ_0 cannot depend on y.
3. The expansion of the Poisson bracket to order μ being zero leads to

$$-\sum \frac{\partial \Phi_0}{\partial x_i} \frac{\partial F_1}{\partial y_i} + \sum \frac{\partial F_0}{\partial x_i} \frac{\partial \Phi_1}{\partial y_i} = 0.$$

Fourier expansion of F_1 produces Fourier coefficients $B(x)$ and an infinite set of integers m_i corresponding to the combination angles of the y_i; in the same way,

Fourier expansion of Φ_1 produces Fourier coefficients $C(x)$. The relation for the
Poisson bracket to order μ must be valid for all $x \in D$ (§82), so that either

$$B(x) = 0 \quad \text{or} \quad \sum m_i \frac{\partial \Phi_0}{\partial x_i} = 0.$$

In general, B is given, so the second condition has to be satisfied for any system
of integers m_i. This leads to the contradiction that Φ_0 is dependent on F_0, so in
general there cannot be another integral.

The Case of Vanishing Fourier Coefficients $B(x)$ (§83)

Consider the exceptional case of vanishing Fourier coefficients. To focus ideas, we
consider systems with two degrees of freedom; more generally, the case of n degrees
of freedom runs in the same way, but the analysis is somewhat more complicated.
As we have seen, for each system of indices m_1, m_2 we have a coefficient $B(x)$.
Suppose that for certain values of x, we have

$$m_1 \frac{\partial F_0}{\partial x_1} + m_2 \frac{\partial F_0}{\partial x_2} = 0.$$

The corresponding coefficient B will be called *secular*. To prove the preceding
theorem, we have assumed that a coefficient B, which is secular, does not vanish.
   Consider now a class of indices with the property that $m_1/m_2 = $ constant. The
class will be called singular if the corresponding secular coefficients B vanish. If
not, the class is called ordinary. In the last case it can be argued that Φ_0 has to be a
function of F_0, and the nonexistence theorem carries over to this case. In the singular
case, we cannot apply the theorem.

The Case of the Singular Jacobian $|\partial F_0/\partial x| = 0$ (§84)

In a number of mechanics problems, F_0 does not depend explicitly on all the
variables $x_1, x_2, \ldots, x_n$. Suppose, for instance, that $F_0(x)$ depends on x_1, x_2 and
not on the other $(n - 2)$ variables x_i. As before, we can again introduce ordinary
and singular classes of coefficients, but now for $2n - 4$ equations. In the ordinary
case, there exists no other independent integral besides the energy F, while in the
singular case, there exist a number of distinct first integrals.

Applications to the Planar Gravitational Three-Body Problem (§85)

1. Consider the planar circular restricted three-body problem as in §9; see Figure 9.2.
   Two of the three masses are $1 - \mu, \mu$, with μ a small parameter, whereas the third

mass is so small that its motion does not affect the motion of the other two bodies. The motion of the third body under the influence of the first two bodies moving in circular orbits is described by the system with two degrees of freedom derived from the energy integral

$$F(x_1, x_2, y_1, y_2) = F_0 + \mu F_1 + \cdots = \frac{1}{2x_1^2} + x_2 + \mu F_1 + \cdots .$$

We have for the Jacobian determinant $|\partial F_0/\partial x| = 0$, but we can use F_0^2 instead (see §43). We have that F_1 is periodic in y_1, y_2, and its perturbation expansion produces secular terms. We conclude that no second independent integral exists that is periodic in y_1, y_2.

2. Consider the planar three-body problem with primary mass m_1 and the two other masses of order μ. If $\mu = 0$, we have again Keplerian orbits. The system has six degrees of freedom, but using the constant motion of the centre of gravity and an additional reduction based on the angular momentum integral, it can be reduced to a system with three degrees of freedom. Analysis of the secular part of the perturbation function shows that this does not vanish, so that no independent integral in the variables of the reduced system exists. A separate analysis is necessary to show that this also carries through for the original system with six degrees of freedom in the sense that in this system we have energy and angular momentum integrals only.

3. For the general planar three-body problem with two small masses we have six degrees of freedom, and we can show in the same way that there exist no integrals independent of the energy and angular momentum. This result is more general than the result of Bruns [Bruns 1888], who proves the nonexistence only of algebraic integrals , while we admit here transcendental integrals. On the other hand, it is more restricted, since Bruns does not assume smallness of the two masses.

The chapter concludes with a discussion of some classical problems of mechanics such as that of rotating solids whereby one looks for an integral of motion. In addition, there are remarks on the case that an integral is not holomorphic in the small parameter, for instance when we have an expansion in fractional powers of the small parameter. Finally, it is noted that for the general (spatial) three-body problem, the same line of reasoning can be followed, for instance by expanding the perturbation function F_1 of the energy integral for small eccentricities and small inclinations.

Comments on Chapter 5

This is the fundamental chapter on the nonexistence of uniform integrals for near-integrable Hamiltonian systems with n degrees of freedom with a small parameter μ. The chapter is conceptually important, since the treatment is much more general than in the prize essay [Poincaré 1890b], which uses series expansions for the three-body problem. A fundamental conclusion is that without additional assumptions, we will not find additional independent first integrals besides the energy F.

A number of exceptional cases can also be dealt with, and a number of three-body problems are analysed with regard to integrability. A problem is again that the Jacobian determinant of F_0 in the expansion $F = F_0(x) + \mu F_1(x, y) + \mu^2 F_2(x, y) + \cdots$ can be singular.

The results are very general, for the analysis takes the completely new perspective that integrability of a Hamiltonian system or even the existence of one additional integral besides the energy is exceptional. There are many problems remaining:

1. In 1923, Fermi raised the question whether a particular $(2n - 1)$-dimensional manifold could exist that depends smoothly on the small parameter μ (of course independent of the energy). His answer was negative for Hamiltonian systems with three or more degrees of freedom; see [Fermi 1923], and see also the discussion in [Benettin et al. 1982], which contains an extension.

2. Up till now, the actual construction by Poincaré has been used mostly for polynomial expansions, producing sometimes formal integrals; more general expansions are rare but necessary.

3. In the case of these formal expansions, the issue is obscured by the fact that nearly all computations have been for systems with two degrees of freedom. In this case, there is no obstruction to computing as many terms of the series as one wishes, since the nonintegrability in a neighbourhood of a stable equilibrium shows up in a measure that is smaller than any power of the small parameter μ; see the chapter on Hamiltonian systems in [Sanders et al. 2007].

4. A technical difficulty is that in an expansion to the small parameter, the system may be integrable to a certain power of μ but not beyond it. So the system is still nonintegrable, but the measure of nonintegrability is restricted by this result; see again [Sanders et al. 2007].

5. It is well known that certain symmetries produce integrals, for instance spherical and axial symmetry. Other symmetries, such as, for instance, discrete (reflection) symmetry, produce other effects. Sometimes, a dynamical system has hidden symmetries that show up in a complicated way. There are many open problems here.

A remarkable extension of our understanding of integrability was supplied by KAM theory (KAM from A. N. Kolmogorov, V. I. Arnold, and J. K. Moser). Its message is that generically for a small perturbation of an integrable (nondegenerate) Hamiltonian system, a positive measure of invariant tori around the periodic solutions will survive in phase space. For details and references see [Broer 2004] or [Kozlov 1996].

As is visualized in Figure 9.6 by a sketch and in Figure 9.11 for a genuine dynamical system, this also applies to area- or measure-preserving maps producing closed invariant curves around stable fixed points. One should realize that in increasing the number of degrees of freedom n, the part played by the invariant

Fig. 9.6 In a cross section of a family of tori embedded in an energy manifold of a Hamiltonian system with two degrees of freedom, one finds stable and unstable fixed points. According to the KAM theorem, the tori form a set of positive measure. The unstable solutions have nearby chaotic motion that is associated with the doubly asymptotic solutions described in Chapter 33

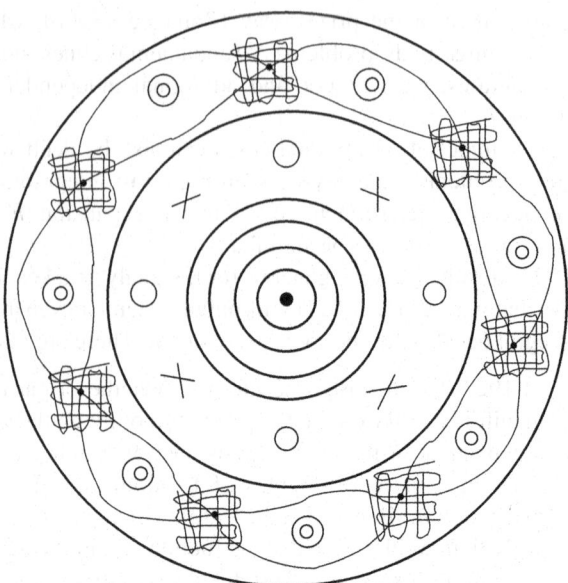

KAM tori decreases. If $n = 2$, phase space is four-dimensional, the energy manifold three-dimensional, and the two-dimensional KAM tori foliate the energy manifold with possible chaotic motion trapped between the tori. If $n = 3$, phase space is six-dimensional, the energy manifold five-dimensional, and the KAM tori on the energy manifold are three-dimensional, since we have three free angles. So the KAM tori do not separate the flow on the energy manifold, and this freedom of the chaotic orbits increases with n (the energy manifold is $(2n - 1)$-dimensional; the tori are n-dimensional).

Chapter 6: Approximation of the Perturbation Function

In the equations of dynamics we have an expansion with respect to a small parameter μ of the energy function (Hamiltonian) $F = F_0 + \mu F_1 + \mu^2 F_2 + \cdots$. We can solve the equations for $\mu = 0$. The perturbation function F_1 contains actions and angles, two angles in the case of two degrees of freedom; F_1 can be expanded in an infinite Fourier series with respect to the angles, where the size of the coefficients determines the relative importance of the expansion terms. Darboux developed a method to determine the relative size of the Fourier terms in the case of one angle based on the analytic properties of the perturbation function. Poincaré extends the method of Darboux to the case of more angles.

First we assume that F_1 consists of terms depending on one angle only (we can apply the method of Darboux directly to these terms) plus terms of mixed form containing more angles. This mixed part will be called the *principal part of the perturbation function* F_1^0. The Fourier coefficients are determined by contour integrals in the complex plane of the form

$$\frac{1}{2\pi i} \int F_1^0 \, dt.$$

Using a Laurent series for the principal part F_1^0, the size of the integral is determined by the number and nature of the singularities of the analytic continuation of F_1^0. We can apply this analysis to problems of dynamics and explicitly to various models of the three-body problem.

An additional aspect is that this analysis helps us to refine results regarding the nonexistence of integrals as obtained in the preceding chapter.

Comments on Chapter 6

Study of the perturbation function is concerned with the Fourier expansion of the term F_1 in the expansion of the Hamiltonian F. The method of Darboux regarding Fourier expansions enables us to determine the relative size of the terms; this is important for applications. This analysis is in itself general but is applied to the three-body problem.

Chapter 7: Asymptotic Solutions

To start with, the term "asymptotic" is used in the sense of behaviour for $t \to \infty$; this changes later on. Consider a general system of the form

$$\frac{dx}{dt} = X(x, t)$$

that is 2π-periodic in t and can be expanded in a power series with respect to x. Suppose that $x^0(t)$ is a particular 2π-periodic solution. Then putting $x = x^0(t) + \xi$, we obtain in the usual way an equation for ξ; the right-hand side is 2π-periodic in t and can be expanded in a power series with respect to ξ. Linearization of the equation for ξ produces a linear periodic system with a solution of the (Floquet) form

$$e^{\alpha t} \Phi(t)$$

with α the characteristic exponents and $\Phi(t)$ a 2π-periodic matrix. We transform

$$\xi = \Phi(t)\eta$$

to obtain an equation for η with right-hand side that is 2π-periodic in t and can be expanded in a power series with respect to η. We can solve the system for η by a recurrent process leading to powers of terms of the form $A_i(t)e^{\alpha_i t}$, where the $A_i(t)$ are 2π-periodic coefficients and the α_i are the characteristic exponents. A problem may arise because in the expansions we find small denominators containing combinations of the characteristic exponents.

Suppose that we have a nonresonance condition guaranteeing that the small denominators do not occur. Solutions with positive α_i will tend to zero as t tends to $-\infty$. In the original system they correspond to solutions that tend towards the periodic solution $x_0(t)$ as t tends to $-\infty$. If $\alpha_i < 0$, they tend to $x_0(t)$ as t tends to $+\infty$. In both cases we will call these solutions *asymptotic solutions*.

We have a more specific analysis when the vector field is restricted to the equations of dynamics (derived from an autonomous Hamiltonian) with small positive parameter μ. As we have seen in the discussion of Chapter 4, the characteristic exponents can be expanded in powers of $\sqrt{\mu}$. The expansion is given explicitly in §108, where it is shown that no negative powers of $\sqrt{\mu}$ arise. The series has coefficients with parameters, and unfortunately, the structure of these terms is such that the series generally diverge. Does this observation dispose of the whole calculation? Such is, in fact, not the case. Poincaré begins with a simple illustration. For small positive μ, consider (§109) the series

$$F(x,\mu) = \sum_{n=1}^{\infty} \frac{w^n}{1+n\mu}.$$

The series converges uniformly for $|w| \leq q < 1$. To obtain an expansion with respect to μ, we calculate the derivatives at $\mu = 0$:

$$\frac{\partial F}{\partial \mu} = -\sum_{n=1}^{\infty} nw^n, \quad \frac{\partial^2 F}{\partial \mu^2} = \sum_{n=1}^{\infty} 2n^2 w^n, \quad \frac{\partial^p F}{\partial \mu^p} = \sum_{n=1}^{\infty} (-1)^p p! n^p w^n.$$

The expansion of F with respect to μ will have a divergent series of the form

$$\sum_{n=1}^{\infty} (-n)^p w^n \mu^p.$$

This seems useless until one realizes that one can drop all the terms with $n > p$, call the collection of remaining terms $\Phi_p(w,\mu)$, and note that

$$\lim_{\mu \to 0} \frac{F(w,\mu) - \Phi_p(w,\mu)}{\mu^p} = 0.$$

So for μ small enough, $\Phi_p(w,\mu)$ is an approximation of $F(w,\mu)$.

In the same way, we will use the expansions obtained for the characteristic exponents and the corresponding quantities to find valid approximations. It makes sense to rephrase the equations and the corresponding expansion to obtain estimates for the errors. In this way, the last sections show that the approximations obtained earlier are asymptotic approximations analogous to the series of Stirling.

Volume 2: 1893

In the introduction, Poincaré mentions that in the methods of scientists such as Newcomb, Gyldén, Lindstedt, and Bohlin, the emphasis for the study of differential equations is on series expansions and the avoidance of secular terms that arise through the use of recurrence procedures. It turns out that this is possible by various methods and that one can obtain rigorous error estimates. However, a fundamental problem that remains is the presence of small denominators in the expansions that may destroy the error estimates. In this respect, Gyldén's method seems mathematically the most sound.

In the notation, μ will again be a small positive parameter, F the total energy of a dynamical system that can be expanded in powers of μ. In examples involving the three-body problem, the primary mass, here called m_1, will be dominant. The other masses are of size μ.

Chapter 8: Formal Aspects of Convergence

Consider two series with respectively the general terms

$$\frac{1000^n}{n!} \quad \text{and} \quad \frac{n!}{1000^n}.$$

Mathematicians will conclude that the first series converges, while the second one diverges. Physicists, astronomers, and engineers will use only the first few terms and will conclude the opposite about the second series. The application-oriented scientists are right in the sense that we can attribute a meaning to the second series, analogous to the series of Stirling, as follows.

Consider a function $\phi(x, \mu)$ and a divergent series of the form

$$f_0 + \mu f_1 + \cdots + \mu^p f_p + \cdots$$

with coefficients $f_0, f_1, \ldots$ constant or depending on x and μ. Put

$$\phi_p = f_0 + \mu f_1 + \cdots + \mu^p f_p.$$

The divergent series will approximate the function $\phi(x, \mu)$ asymptotically if the coefficients $f_0, f_1, \ldots$ are bounded by constants independent of μ and

$$\lim_{\mu \to 0} \frac{\phi - \phi_p}{\mu^p} = 0.$$

For illustration, one can consider the n-dimensional equation

$$\dot{x} = X(x, t),$$

where X is an analytic function of x and t that can be expanded in increasing powers of μ. Substitution of a series in increasing powers of μ with functions of t as coefficients produces a formal, in general divergent, series satisfying the differential equation. If the coefficients still depend on μ, they are supposed to have a convergent expansion with respect to μ. Then it is shown that the formal series thus obtained, provided it obeys certain uniformity requirements, represents an asymptotic approximation of the solution.

Comments on Chapters 7 and 8

The observations on asymptotic approximations start with asymptotics in the usual sense (behaviour at infinity) and then proceed to discuss approximations using divergent series expansions. This was quite controversial at the time, since following the use of formal series expansions in the eighteenth century, rigorous criteria for convergence had been developed in the nineteenth. One of the mathematicians who played a prominent part in this was Cauchy (1789–1857). To accept again divergent or asymptotic series as a tool went against the common mathematical sense of the time. An asymptotic series as an approximation of a solution need not converge, and if it converges, it need not converge to the solution. To illustrate the last point, consider the function

$$f(x) = \sum_{n=0}^{\infty} \mu^n c_n(x) + \sum_{n=1}^{\infty} e^{-(\mu+x)/\mu^{2n}},$$

with μ a small positive parameter, $0 \le x \le 1$; both series on the right-hand side converge on the interval $[0, 1]$. Any partial sum of the first series $S_N = \sum_{n=0}^{N} \mu^n c_n(x)$ represents an asymptotic approximation of $f(x)$ with $f(x) - S_N = O(\mu^{N+1})$ as $\mu \to 0$.

Poincaré published these ideas first in 1886 [Poincaré 1886]; in the discussion of Chapter 8, the formulation is not very precise, a fact that made mathematicians around 1900 probably even more suspicious. In the same year, the Dutch mathematician Thomas J. Stieltjes (1856–1894) defended his thesis at the Sorbonne under the supervision of Hermite, formulating the same ideas as "semiconvergent series"; see [Stieltjes 1886]. Both Stieltjes and Poincaré refer to an example given by Stirling, but remarkably enough, not to each other (Stieltjes chose for his oral examination Poincaré's work on rotating fluid masses, but that is a different topic). Asymptotic approximations in the sense of Chapter 8 are now fully accepted and frequently used concepts in mathematics.

Divergence of Series in Hamiltonian Systems

The divergence of series in Hamiltonian systems is a famous and difficult topic that has seen many papers and also was the cause of much confusion in the scientific literature. Most mathematical physicists were aware that integrability and convergence in dynamical systems were difficult subjects, but most of them did not realize that there were fundamental obstructions. The doubly asymptotic solutions, later described by Poincaré in Chapter 33—in modern language, the presence of transversal homoclinic points—are at the basis of the divergence of the series in Hamiltonian systems with at least two degrees of freedom. The dynamics is illustrated in Figures 9.6 and 9.11. The modern theory is discussed in [Broer 2004, Arnold 1978, Arnold et al. 1988, Kozlov 1996].

Chapter 9: The Methods of Newcomb and Lindstedt

The method of Lindstedt (1854–1939) is concerned with obtaining series expansions for solutions of the equation

$$\ddot{x} + n^2 x = \mu\phi(x, t),$$

where $\phi(x, t)$ can be expanded in a power series of x and is periodic in t. Lindstedt also extends this to a system of two coupled harmonic oscillators. The method is formal, i.e., the approximate character of the expansion is not clear. Moreover, Lindstedt had doubts about the possibility of continuing the expansion to arbitrary order. Earlier, in 1874, Newcomb (1835–1909) developed a similar method for application to the three-body problem.

Poincaré presented a very general form of the method, based on the implicit function theorem, and in addition gave a rigorous justification. In §125 we start again (as in §13) with the general equations of dynamics

$$\dot{x} = \frac{\partial F}{\partial y}, \quad \dot{y} = -\frac{\partial F}{\partial x},$$

where $F = F_0 + \mu F_1 + \mu^2 F_2 + \cdots$ and F is 2π-periodic in y. We will look for a power series expansion in μ of x and y with time-dependent coefficients $x^i(t), y^i(t)$ that can be expanded in a Fourier series. The procedure to determine this formal series will be different from Lindstedt's procedure. We will use a series in terms of y and μ, derived from the Jacobi generating function of the form

$$S(y) = S_0(y) + \mu S_1(y) + \mu^2 S_2 + \cdots,$$

where the coefficients S_k consist of linear terms in y plus terms that are 2π-periodic. The series has to satisfy formally the partial differential equation

$$F\left(\frac{\partial S}{\partial y_1}, \ldots, \frac{\partial S}{\partial y_n}, y_1, \ldots, y_n\right) = \text{constant.}$$

Comparing the coefficients of equal powers of μ, we obtain a recurrent system of equations for S_k. In addition, we require the system to remain canonical, and thus we obtain expansions that are either convergent or divergent approximations in the sense of the preceding chapter.

One can formulate the series in different ways as discussed in §§126–127. A novel aspect is that one can also expand the angles, or just time in simple problems, with respect to μ. The averages of the expansion terms have to satisfy conditions depending on the specific application. Certain degeneracies (for instance the Jacobian determinant of F_0 vanishing), as in the example of the three-body problem in §9, can be transformed away, but sometimes this poses a more serious problem.

Newcomb, in his series approximations for the three-body problem, starts with Lagrange variation of constants. This is in essence an equivalent way of obtaining the right series expansions for such applications.

Comments on Chapter 9

The ingenious method of Newcomb and Lindstedt of obtaining formal approximations of solutions of differential equations is raised by Poincaré to a much higher level through the development of a method for the general equations of dynamics. The method enables us to avoid secular terms in the expansion with respect to a small parameter. The series, however, is generally divergent; this is discussed in more detail in Chapter 13.

Chapter 10: Secular Variations

Notwithstanding its general title, this short chapter is concerned with perturbations in the three-body problem. In certain models, the perturbation function is reduced because the angles vary, but the actions vary at a higher order of μ. In addition, the calculations can be improved by also expanding with respect to eccentricity and inclination. In this way, one recovers the stability results for the solar system of Lagrange and Laplace, who used elimination of secular terms by averaging, an early form of normalization. But of course these results represent only approximations and do not settle the stability question.

Chapter 11: Application to the Three-Body Problem

As noted before, the three-body problem has the special difficulty that in the expansion of the Hamiltonian $F = F_0 + \mu F_1 + \mu^2 F_2 + \cdots$, F_0 is independent of some of the positional x variables. This problem is considered in a much more general context. Suppose to start with that we have three degrees of freedom and that $F_0 = F_0(x_1, x_2)$, and F_1 depends on the six x, y variables and is 2π-periodic in the

y variables. The function $R = R(x_1, x_2, x_3, y_3)$ equals the function F_1 averaged over the variables y_1 and y_2. Following the ideas of §125, we can construct an expansion for a generating function S that will be used for a suitable canonical transformation. The function R plays an essential part in this construction; the expansion for its construction is also based on the standard normal-form nonresonance condition. This guarantees convergence of the expansion. An extension to more than three degrees of freedom is straightforward.

Application to the three-body problem involves the usual calculational difficulties. As indicated in Volume 1, distinguishing the cases of small eccentricities and small inclinations is useful.

Chapter 12: Application to Orbital Calculations

An interesting difficulty arises in practice. Consider as an example the Hamiltonian

$$F = \Lambda + \mu \left(\sqrt{\Omega} \cos(\omega + \lambda) + A\Omega \right),$$

where A is a constant, and Λ, λ, and Ω, ω are conjugate variables. The corresponding equations of motion can be integrated in terms of elementary functions, but it is instructive to follow the series-expansion scheme of the preceding chapter. The transformations involve a variable V that depends on the orbital elements, but the expansion turns out to contain terms of the form μ^2/V. It is clear that if $V < \mu^2$, then the expansion ceases to be valid.

The problem is analogous for the three-body problem. The small parameter μ is produced by the smallness of two of the masses. However, the expansion contains, apart from terms with powers of μ, also terms with powers of eccentricities e and $1/e$. The simple example above suggests the right change of variables, and a suitable shift of the coordinate system modifies and saves the expansion. Using the fact that we are discussing type-one solutions as introduced in Chapter 3 (Volume 1), we can also use an appropriate change of variables to save the expansion for the three-body problem.

Comments on Chapters 10–12

Chapters 10 through 12 are concerned with orbital calculations, in particular for the three-body problem. The case in which the Jacobian determinant of F_0 is singular is discussed in detail. The conclusion is that the stability of the solar system obtained by series expansions is formal; the stability question itself remains open even today, except for the rather unrealistic model of the restricted planar circular case.

Chapter 13: The Divergence of the Lindstedt Series

Consider again the Hamiltonian equations of motion with small parameter μ. We have shown that the equations are satisfied by a formal series of the form

$$x = x^0 + \mu x^1 + \mu^2 \ldots,$$

$$y = w + \mu y^1 + \mu^2 \ldots,$$

where w is a constant n-vector; x, y are n-vectors, periodic in $w = \omega t + \Pi$, where $\omega = (\omega_1, \omega_2, \ldots, \omega_n)$, and Π are constant n-vectors. For the solutions x and y, the formal series is of the (Fourier) form

$$A_0 + \sum_i A_i \cos(m_1 w_1 + m_2 w_2 + \cdots + m_n w_n + h).$$

(Poincaré uses n_i instead of ω_i, but we want to avoid confusion with the number of degrees of freedom.) Earlier, we saw that the coefficients contain terms of the form

$$\frac{B_i}{m_1 \omega_1 + m_2 \omega_2 + \cdots + m_n \omega_n}.$$

If the frequencies are rationally dependent, the denominators may vanish, but if they are independent over the rationals, the denominators may still become arbitrarily small. We conclude that depending on the coefficients B_i (or A_i), the series will generally diverge, but there will be cases in which it converges. It is noted that in practice, it may occur that the Hamiltonian function F and in particular its first expansion terms have a Fourier series with terms that rapidly diminish in size. In such a case, we can shift the infinite tail of the series to higher order in μ, resulting in an approximating expansion that is convergent even though the complete series is not.

One question to be answered in §148 is whether it is possible that the series solution converges uniformly for all values of μ in a certain domain. This convergence would lead to the existence of n different integrals of motion and to $2n$ characteristic exponents of a periodic solution being zero. The answer is that in general, this is not the case.

A second question is whether for μ small enough and for x_0 conveniently chosen, the series solution can be made uniformly convergent. In §149 it is argued that this is not very probable, but the question was not settled conclusively in Poincaré's time.

A final warning is added: representation of a solution in purely trigonometric terms implies many assumptions about the solutions. Without more a priori knowledge of the solution than we have, its use is restricted.

Comments on Chapter 13

Regarding the divergence of the Lindstedt series, the question of the small denominators becomes more explicit. This raises the difficult question whether locally in the variables and for small values of μ, the series can be convergent. Poincaré argues against this with sound reasoning for the general Hamiltonian equations of dynamics, but convergence remains possible in special cases. The part played by resonance in the calculations is now well established in the theory of normal forms of dynamical systems. The term "Poincaré domain" is used to indicate a certain domain in parameter space distinguished from the so-called Siegel domain (see [Verhulst 2000]).

One cannot expect convergence for Hamiltonian systems. The discussion is tied in with the appearance of transverse homoclinic points in maps characterizing the dynamics as referred to in the comments on Chapters 7 and 8. We have no convergence of the perturbation series for solutions with "general" initial conditions in the presence of an infinite number of periodic, quasiperiodic, and chaotic solutions. On the other hand, if we have a priori knowledge of the existence and location of a periodic solution, we have a strong case for the construction of a convergent approximating series. The argument becomes even stronger when we go beyond time-independent Hamiltonian systems to find attracting or repelling periodic solutions. Continuation by the Poincaré–Lindstedt method can in such cases produce existence and a convergent approximation of these solutions; for an introduction, see [Hale 1969] or [Verhulst 2000]. The classical example is the approximation of the periodic solution of the Van der Pol equation.

A fundamental step forward was represented by the results of Siegel in 1942 and the KAM theorem after 1954; see the survey [Broer 2004] or [Arnold et al. 1988]. In these papers, it was shown that for certain initial conditions, in fact an infinite number of them, certain series expansions are convergent and will correspond to invariant tori around stable periodic solutions. Such results are tied in with the question of the nonexistence of first integrals as discussed in Chapter 5 and the comments there.

Chapters 14–15: The Direct Calculation of the Series

Once the theory of the preceding chapters has been accepted, the calculation of the perturbation series can be done efficiently, depending on the type of problem at hand. Canonical transformations and averaging techniques play a part, but in the application to the three-body problem starting in §152, there turn out to be many special cases. In §157 of Chapter 13 there is an important conclusion:

> Thus are the series that one gets by the calculational procedures explained in the preceding chapters. It was Newcomb who was the first to have this idea and who has discovered the main properties. The series are divergent, but if one stops the expansion in time, i.e., before having encountered small divisors, they represent the solutions with a very good approximation.

Fig. 9.7 The restricted plane three-body problem in relative coordinates with dominant central mass μ. The mass of the second planet is so small that it does not influence the two Keplerian orbits of the other masses

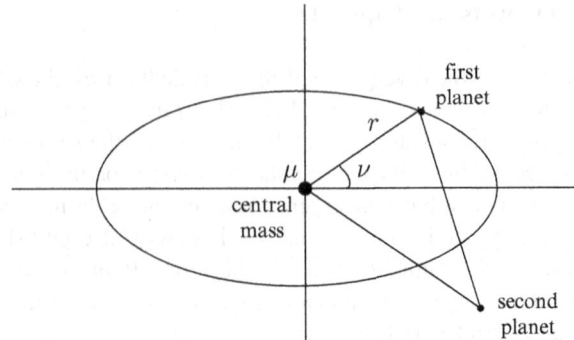

Poincaré adds that the results can also be used in another way. One can use the approximations as new variables that will produce by variation of constants in the original equations a system with extremely slowly varying coefficients.

There is considerable freedom in choosing the transformations and also the averages employed in the expansions. In Chapter 15, a number of cases are discussed to illustrate various procedures.

Comments on Chapters 14–15

These chapters are concerned with the efficient calculation of orbits; there are many different cases. In Poincaré's time, series expansion was the most important quantitative method available, since numerical calculations were still too laborious. At present, numerical methods have not replaced series expansions. Rather, their function is supplementary, adding essential numeric information and visualization.

It is of interest to note that one can use approximations as new variables to formulate higher-order perturbation problems. This idea was not much used in the literature, an exception being the study of higher-order resonances of Hamiltonian systems with two degrees of freedom (see [Sanders et al. 2007]).

Chapter 16: The Methods of Gyldén

The methods of Gyldén are partly related to techniques discussed before and are partly concerned with cases that are not applicable to the Lindstedt method of Chapter 9 and the direct methods of Chapter 15. All of them were originally formulated with the purpose of studying the three-body problem.

A certain independent variable ν_0 plays a prominent part. Consider three bodies moving in a plane. Suppose that μ is the mass of a central body placed at the origin of the coordinate system. Let r, ν be the polar coordinates of one of the two planets revolving around the central mass; see Figure 9.7. The equations of motion are of the form

$$\frac{d}{dt}\left(r\frac{dv}{dt}\right) = \frac{\partial\Omega}{\partial v}, \quad \frac{d^2r}{dt^2} - r\left(\frac{dv}{dt}\right)^2 + \frac{\mu}{r^2} = \frac{\partial\Omega}{\partial v},$$

where $\Omega(r, v)$ is the perturbation function arising from the second planet. In the unperturbed case, $\Omega = 0$, one recovers the Kepler problem, and putting $r = -1/u$, one can derive the equation

$$\frac{d^2u}{dv^2} + u + \frac{\mu}{c} = 0,$$

with c^2 the constant angular momentum. Inspired by this, Gyldén introduces the independent variable v_0 by

$$r^2\frac{dv_0}{dt} = \sqrt{c_0},$$

with c_0 a new constant. The equation for u can be written as

$$\frac{d^2u}{dv_0^2} + u + \frac{\mu}{c_0} = \frac{r^2}{c_0}\frac{\partial\Omega}{\partial r} + u\left(1 - \left(\frac{dv}{dv_0}\right)^2\right).$$

The terms on the right-hand side are small. A similar equation can be derived for the motion of the second planet using an independent variable v_0. In this way, our problems are formulated as perturbed linear equations. However, we need the definitions of the independent variables to relate them to time t. This poses a calculational complication, since this relation will change at each level of approximation. The perturbation scheme based on these equations will involve second-order linear and nonlinear equations with periodic coefficients. Using averaged terms in the equations, Gyldén obtains so-called intermediate orbits that can be used as a beginning for approximations.

Poincaré notes that the independent variable v_0 can be replaced by other variables with similar results.

Comments on Chapter 16

The chapter discusses the methods of Gyldén for series expansions in the three-body problem. It was noted by several contemporaries of Poincaré that when reading scientific papers, he looked at the problem formulation, then at the final result, and next constructed the reasoning in between by himself. In the case of the work of Gyldén, this might have helped, since even Gyldén's student Hugo Buchholz found his papers hard to read (see the discussion in [Poincaré 1999], letters 80–84). The discussion by Poincaré in Chapter 16 was rather delicate, since Gyldén had complained to the prize committee—see [Barrow-Green 1997]—that the prize essay [Poincaré 1890b] contained results that he had obtained much earlier. This claim was certainly not correct, since Poincaré proved in addition that some of Gyldén's expansions were divergent, while all of them were merely formal.

The methods are summarized in the chapter and are not referred to again. In §169 it is mentioned that Gyldén's equations are not canonical but can be put in canonical form. Bohlin's approach (Chapters 19–20) is canonical.

Chapter 17: The Case of Linear Equations

This chapter is devoted to linear equations and various methods. Consider one of Gyldén's equations written as

$$\ddot{x} = x \left(-q^2 + q_1 \cos 2t \right) \tag{9.4}$$

with constants q, q_1. With the characteristic exponents $h, -h$, the independent solutions can be written as

$$e^{iht\phi_1(t)}, \quad e^{-iht\phi_2(t)},$$

with $\phi_1(t)$, $\phi_2(t)$ π-periodic. Poincaré notes that he proved in a paper (*Acta Math.* 4, p. 212) that the solutions of the linear equation (9.4) can be expanded in a series of increasing powers of q^2 and q_1; in fact, the paper discusses far more general linear equations. Using this theorem and the formulation of the solutions in terms of characteristic exponents and periodic functions $\phi_1(t)$ and $\phi_2(t)$, one can explicitly determine the expansion coefficients. Interesting cases arise whenever $q = 2n$ (even) or $q = 2n + 1$ (odd). This leads to approximations of the characteristic exponents and the determination whether for given q, q_1, they are purely imaginary. A figure in §179 shows stability and instability of the equilibrium $(0, 0)$ of equation (9.4) by what is now usually called the "instability tongues of the Mathieu equation."

The Method of Jacobi (§181)

One can put (9.4) in canonical form using action-angle variables and solving the PDE for the generating function. It is then possible to apply the approximation method outlined in §125, producing recurrent relations for the expansion coefficients.

The Method of Gyldén (§182)

Gyldén applies a theorem of Picard for linear equations with doubly periodic coefficients. Picard shows that if the general solution has no worse singularities than poles, the solutions change by a constant factor if one increases t by a period. In fact, Hermite applied this already to the Lamé equation. Gyldén uses the theorem to approximate the solutions of (9.4).

The Method of Bruns (§183)

Bruns transforms (9.4) into a first-order equation for z by putting

$$x(t) = e^{\int z(t)dt}.$$

The solutions of the first-order equation can be expanded in powers of q_1; it is easy to see that if q is not an integer, then the expansion terms are periodic in t.

The Method of Lindstedt (§184)

Lindstedt's method should not be confused with the Poincaré–Lindstedt small-parameter method. Consider an even solution of (9.4) and expand this in a cosine series. For this we need the characteristic exponent and the coefficients. We can consider as well the equation (9.4) with added right-hand side $\beta \cos \lambda t$. It has a solution of the form

$$x = \sum B_n \cos(\lambda + 2n)t.$$

Substituting the series in the inhomogeneous equation, we obtain

$$B_n \left(q^2 - (\lambda + 2n)^2 \right) = \frac{q_1}{2}(B_{n-1} + B_{n+1}), \quad n \neq 0,$$

$$B_0 \left(q^2 - \lambda^2 \right) = \frac{q_1}{2}(B_{-1} + B_{+1}) + \beta.$$

Putting $\lambda = h$ and $\beta = 0$ produces the above-mentioned expansion for (9.4). Introducing

$$\alpha_n = \frac{B_n}{B_{n-1}} \quad \text{for} \quad n > 0, \quad \alpha_n = \frac{B_n}{B_{n+1}} \quad \text{for} \quad n < 0,$$

we can derive continued fractions for the α_n. This leads to recurrence relations that produce convergence to determine B_n and λ. Afterwards, we can insert $\lambda = h$ and $\beta = 0$ to obtain the solution of (9.4). In general, the determination of the characteristic exponent h in this way is more complicated than by the previous methods.

The Method of Hill (§185–189)

The equations obtained by expanding in a series are linear, and Hill proposed to treat them with the methods of linear algebra. This involves matrices and determinants with an infinite number of entries. Poincaré summarizes in §185 his own results on this topic.

Consider a square matrix with an infinite number of columns and rows with all diagonal elements equal to 1. Let Δ_n denote the determinant of the matrix a_{jk}, $j, k = 1, \ldots, n$, formed by the first n columns and rows. If certain products of the entries a_{jk} converge for $n \to \infty$, we have that Δ_n converges for $n \to \infty$. Absolute convergence guarantees that we can interchange columns and rows without affecting the limit of the Δ_n.

Hill uses infinite determinants to compute the motion of the perigee of the Moon. The entries of the determinants have to satisfy convergence properties for which a theorem of Hadamard can be useful.

Comments on Chapter 17

This chapter is concerned with expansion for (mainly) the Mathieu equation, which appears as a variational equation to describe the solutions near a periodic solution. This leads to the instability tongues for certain resonance values. Hill and Poincaré introduced matrices with an infinite number of rows and columns to study this equation; this important idea was taken up later by Fredholm (1866–1927) for more general operator equations.

Chapter 18: The Case of Nonlinear Equations

It can be necessary to obtain approximations of the solutions of the equation

$$\ddot{x} + x \left(q^2 - q_1 \cos 2t\right) = \alpha \phi(x, t),$$

where α is a small parameter and $\phi(x, t)$ consists of terms of the form

$$Ax^p \cos \lambda t + \mu,$$

with p an integer and A, λ, μ constants. An ingenious way to provide more freedom in the expansions is to write the equation as

$$\ddot{x} + x \left(q^2 + \beta + (-q_1 + \gamma) \cos 2t\right) = \beta x + \gamma x \cos 2t + \alpha \phi(x, t).$$

Here β and γ are small parameters. The first-order approximation $\xi(t)$ of a perturbation series for the nonlinear equation is obtained from the preceding chapter by putting $\beta = \gamma = 0$ and $\phi = \phi(0, t)$. For the next approximation we put $\phi = \phi(\xi(t), t)$, and we choose $\beta = \beta_2, \gamma = \gamma_2$ such that the solutions of

$$\ddot{x} + x \left(q^2 + \beta_2 + (-q_1 + \gamma_2) \cos 2t\right) = \beta_2 \xi(t) + \gamma_2 \xi(t) \cos 2t + \alpha \phi(\xi(t), t)$$

contain no secular terms. The equations obtained in this way are linear, and the procedure can be continued to higher order.

The procedure can be formulated more generally for Hamiltonian systems of the usual form

$$F = F_0 + \mu F_1 + \mu^2 F_2 + \mu^3 F_3 + \cdots$$

with μ a small parameter, but this time $F_0 = F(x, y)$, where F_0 is 2π-periodic in the y components (angles). One supposes that the system for $\mu = 0$ can be integrated. At each successive step of approximation, small modifications of both x and y are introduced.

The nonlinear second-order equation formulated in this chapter can be treated in this way, as can as well equations of the form

$$\ddot{x} + f(x) = \mu \phi(x, t).$$

Again one can conclude that Newcomb's method works well when there are no resonances (commensurabilities in the frequencies); the modifications by Gyldén can be applied in the case of resonance.

An important generalization is given in §198, where the possibility of approximating quasiperiodic solutions is considered for the equation

$$\ddot{x} - \alpha x = \mu f(x, t, \mu).$$

The function f depends quasiperiodically on n arguments $\lambda_i t, i = 1, \ldots, n$. A recurrent system yielding a convergent approximation can be obtained in the case that α is arbitrary and there is only one frequency ($n = 1$) and in the case of $\alpha > 0$ and n arbitrary.

Comment on Chapter 18

The nonlinear extension of periodic differential equations in the preceding chapter turns out to be a topic with many different phenomena. Curiously enough, the results of this chapter have not been much exploited in the literature. There is an important extension to quasiperiodic equations in §198.

Chapters 19–20: The Methods of Bohlin

This chapter and the last two of Volume 2 deal with expansions in the case that we have small denominators. The first results were obtained by Delaunay, and §§199–203 are used to explain his method. Suppose again that we have a small-parameter expansion of the Hamiltonian

$$F = F_0 + \mu F_1 + \mu^2 F_2 + \cdots$$

depending on the positions $x_1, \ldots, x_n$, periodically on the combination angle $\chi = m_1 y_1 + \cdots + m_n y_n$, and also periodically on the separate arguments $y_1, \ldots, y_n$; the numbers $m_1, \ldots, m_n$ are integers. Using χ directly in the generating function S instead of the separate angles produces various expansions, depending on the location in phase space (there are constants of integration that depend on the initial values).

An example is presented in §199 to illustrate Delaunay's approach. Consider a system with one degree of freedom with

$$F = x + \mu \cos y.$$

The Jacobi equation for the generating function becomes

$$\left(\frac{dS}{dy} \right)^2 + \mu \cos y = C,$$

with C a constant. The solutions depend on C as follows:

1. $C > |\mu|$. In this case, $\sqrt{C - \mu \cos y}$ is always real, and we can expand

$$S = x^0 y + \sum_n \frac{B_n}{n} \sin ny,$$

with x^0 an arbitrary constant.

2. $-|\mu| < C < |\mu|$. In this case, only values of y are permitted that keep $\sqrt{C - \mu \cos y}$ real. Introducing an auxiliary variable ε by

$$\mu \cos y = C \cos \varepsilon,$$

we deduce that

$$\frac{dS}{d\varepsilon} = \sqrt{\frac{\mu^2 - C^2 \cos^2 \varepsilon}{C}}.$$

Since C^2 is smaller than μ^2, we can expand

$$S = B_0 + \sum_n \frac{B_n}{n} \sin n\varepsilon,$$

yielding S as a function of C and variable ε.

3. The boundary case $C = |\mu|$. Taking for instance $\mu > 0$, we obtain

$$S = -\sqrt{2\mu} \cos \frac{y}{2},$$

which is 4π-periodic. Using S, we obtain solutions expressed in (doubly periodic) elliptic functions.

From the expansions it is clear that if we had started with an expansion with respect to powers of μ, the expansion would not have been convergent for all initial conditions, i.e., for different values of C.

Delaunay's analysis is correct, but it involves many transformations. Both Bohlin and Poincaré modify Delaunay's method to make the procedure more efficient. This leads to expansions determined by recurrent systems with many different cases. As indicated by the simple example of §199, some of the series will be divergent and have to be interpreted in an asymptotic sense as before.

Chapter 21: Extension of the Method of Bohlin

In the three-body problem, we have an additional difficulty in that in the expansion of the Hamiltonian $F = F_0 + \mu F_1 + \mu^2 F_2 + \cdots$, some of the variables may be missing in F_0 (see Chapters 11 and 13). The method of Bohlin can still be applied with subtle modifications. The expansion terms of the generating function S involve various averages. If y_1 corresponds to a resonant angle combination and U is a function periodic in the y_i, then $[U]$ is the function averaged over y_1, and $[[U]]$ is the function U averaged over all y_i, $i = 1, \ldots, n$. For the three-body problem we have various resonances and so various series expansions, whereas rotation and libration of the orbits also play a part. The series can be either convergent or divergent. An example in §225 is used to illustrate this. Consider the Hamiltonian

$$F = -p - q + 2\mu \sin^2 \frac{y}{2} + \mu\varepsilon\phi(y)\cos x;$$

(p, x) and (q, y) are conjugate variables, $\phi(y)$ is 2π-periodic in y, μ and ε are small parameters. The equations of motion are

$$\dot{x} = 1, \quad \dot{p} = -\mu\varepsilon\phi(y)\cos x; \quad \dot{y} = 2q, \quad \dot{q} = \mu\sin y + \mu\varepsilon\phi'(y)\cos x,$$

x being a timelike variable. It makes sense to derive and study the equation

$$\ddot{y} = 2\mu \sin y + 2\mu\varepsilon\phi'(y)\cos x.$$

For Jacobi's equation we obtain

$$\frac{dS}{dx} + \left(\frac{dS}{dy}\right)^2 = 2\mu \sin^2 \frac{y}{2} + \mu\varepsilon\phi(y)\cos x + C,$$

with C a constant. For $\varepsilon = 0$, we can solve the problem, so we expand

$$S = S_0 + S_1\varepsilon + S_2\varepsilon^2 + \cdots, \quad C = C_0 + C_1\varepsilon + C_2\varepsilon^2 + \cdots.$$

For S_0 we obtain

$$S_0 = A_0 x + \sqrt{2\mu} \int \sqrt{h + \sin^2 \frac{y}{2}} \, dy,$$

with A_0, h constants. The analysis runs now as for the example in §199. If $h > 0$, we have the ordinary case, while $h < 0$ corresponds to libration, and $h = 0$ is the boundary case.

We can also compute S_1 and S_2; the higher-order S_i are of the same form. Extracting the corresponding solution series for $h > 0$, it can be explicitly shown that these diverge. Considering the case $h < 0$, nothing much changes except that the boundaries of the respective integrals are not $[0, 2\pi]$ but depend on $-h$. The case $h = 0$ can also be treated completely with the possibility of comparing the various expansion methods and pointing out the intricate problems arising in this example.

Comments on Chapters 19–21

These chapters deal with the method of Bohlin and its extensions, using series expansion in canonical form while avoiding secular terms. This takes care of small denominators, but divergence of the series is to be expected. The calculations conclude the relatively application-oriented Volume 2.

Volume 3: 1899

The third volume contains a number of fundamental ingredients of the modern theory of dynamical systems, for instance the notions of integral invariants (and manifolds), the so-called Poincaré map, periodic solutions, and homoclinic and heteroclinic solutions.

Chapter 22: Integral Invariants

An example of an integral invariant is the volume V of a fluid element in the case of an incompressible fluid. If F_0 describes the fluid element geometrically, we have

$$\int_{F_0} dx \, dy \, dz = V,$$

where we have integration in three-dimensional space. In an incompressible fluid, the volume V of a fluid element is constant. At the same time, we can express the incompressibility by the equation

$$\nabla \cdot v = 0,$$

where v is the velocity field. A similar integral can be written down for a gas where mass is conserved in combination with the continuity equation. More generally, consider the vector field $x = (x_1, x_2, \ldots, x_n)$ generated by the differential equation $\dot{x} = X(x)$. The integral

$$\int_{F_0} A(x) \, d\omega$$

is called an integral invariant of order p if $d\omega$ involves p differentials of $dx_1, \ldots, dx_n$, $A(x)$ is a differentiable function, F_0 a given p-dimensional manifold, and when under the phase flow, the value of the integral is constant.

It is possible that we have an additional condition for the manifold F_0 and the subsequent manifolds generated by the phase flow. For instance, if $p = 1$, we can require F_0 to be a closed curve, or more generally, for $p > 1$, we can require a closed manifold. In this case, we call the integral invariant *relative* to this additional condition. We note that if $p = 1$ and we are considering an integral invariant relative to closed curves, we can apply Stokes's theorem, which enables us to obtain results for potential problems and the relation with exact differentials (§240).

In our discussion of Chapter 4 we formulated the variational equations with respect to a particular solution (often a periodic solution) of a system of differential equations $\dot{x} = X(x)$. The variational equations are of the form $\dot{\xi} = A(t)\xi$. Suppose they have a time-independent integral of the form $F(\xi) = $ constant, where F will be linear and homogeneous in ξ. We can deduce a relation between F and an integral invariant of the original system. Integral invariants can be transformed and combined to obtain new expressions for integral invariants.

Chapter 23: The Formulation of Integral Invariants

In considering an autonomous system of the form $\dot{x} = X(x)$, it is sometimes possible to obtain an integrating factor M. In the case of the (Hamiltonian) equations of dynamics, we have simply $M = 1$. More generally, the expression $\int M dx$ is an integral invariant of the autonomous system.

For the equations of dynamics we can find many integrals, but they will usually be dependent on the known integrals for energy, linear momentum, and angular momentum (integral of areas). This can be shown by calculating the Jacobian. If the potential is homogeneous in its arguments, a new integral may arise. Also, integrals may be time-dependent, an important case being that they are invariant with respect to a closed curve.

The variational equations that are formed with respect to a periodic solution will have characteristic exponents. These equations may have integrals that are linear in the variational variable ξ and algebraic in x, the special periodic solution. To have q independent invariants, q characteristic exponents have to vanish. In §257, the concept of a singular solution is introduced. Suppose we have q independent integrals $I_1, \ldots, I_q$, so we have in general

$$\beta_1 I_1 + \beta_2 I_2 + \cdots + \beta_q I_q \neq 0, \tag{9.5}$$

with the β_i arbitrary constants. If the right-hand side of (9.5) vanishes for a special solution $X(t)$, we call this solution *singular*. A question that will play a part is whether all periodic solutions of the equations of motion are singular.

In §260, four types of invariants are distinguished analytically, leading to four types of integrals with the possibilities for the three-body problem as an illustration.

Chapter 24: The Use of Integral Invariants

The series expansions developed in Volume 2 can be verified using known integral invariants. This also holds for a theorem of Jacobi for periodic solutions of potential problems: the average of the kinetic energy equals the average of the potential energy modulo a constant. The two-body problem is used as an illustration.

Comments on Chapters 22–24

This part formulates and discusses integral invariants; see also [Arnold 1978]. Important is the relationship between the integrability of the variational equations and the existence of integrals of the original system. One can, for instance, prove that to have q linear invariant integrals, each of the nonsingular periodic solutions will have q characteristic exponents equal to zero. These ideas are still waiting to be extended.

Following Poincaré's suggestion, integral invariants can be used as a check on the validity of perturbation expansions. A necessary condition is that expansions satisfy a known integral invariant. Today, such checks also play a part in numerical integration procedures for conservative and reversible systems. Another use of integral invariants in our day is in the interpretation of physical phenomena and the formulation of certain physical systems, for instance in stellar dynamics, that are difficult to handle by traditional statistical mechanics.

Based on Poincaré's treatment of integral invariants, Élie Cartan gave a series of lectures [Cartan 1922] in 1921. He used transformation groups in the spirit of Sophus Lie and Pfaff invariants to develop a general theory with certain physical applications. The nonintegrability result of Chapter 5 is not mentioned; such discussions had to wait until the second half of the twentieth century.

Chapter 25: Integral Invariants and Asymptotic Solutions

The series expansions according to the method of Bohlin lead to asymptotic solutions that can be used to analyse integral invariants. In the case of equations of (Hamiltonian) dynamics with n degrees of freedom, the characteristic exponents count two zeros and $2n - 2$ conjugate complex values; the exponents play a part in the construction of expansions and invariants. Using series expansions explicitly may lead to an asymptotic invariant manifold. Depending on the type of families of periodic solutions, for instance all of them singular in the sense of §257, certain algebraic invariants may exist.

In the case of the restricted three-body problem of §9 (a negligible mass moving in the field of two masses in circular Keplerian orbits), one quadratic integral is known, but the existence of two quadratic invariants is possible in principle. It turns out that there are periodic solutions that are singular, but not all periodic solutions are of this type. This prohibits the existence of a second quadratic integral.

Comments on Chapter 25

The use of integral invariants and asymptotic solutions presents a natural approach to the analysis of solutions and integrals. Poincaré introduces a scaling of the variables $x \mapsto \varepsilon x$, etc., resulting in the scaling of the Hamiltonian $F \mapsto \varepsilon^2 F$. Together with natural small parameters of the problem (the small mass ratio in the three-body problem), this produces asymptotic, nonconvergent expansions. An integral obtained from these expansions will correspond to an asymptotic manifold. In this way, one can obtain approximations of the KAM tori described in the comments to Chapter 5.

For an algebraic integral to exist, the characteristic exponents of the periodic solutions have to obey a particular relation, or the periodic solution has to be singular in the sense of §257. This is an interesting criterion for the possible existence of algebraic integrals. Computationally, the search for algebraic integrals has become much easier nowadays through the use of computer programs for algebraic manipulation. Using slightly different approximation or normalization techniques, the existence of asymptotic integrals has been extensively studied for general Hamiltonian systems with two or three degrees of freedom near stable equilibria; see [Sanders et al. 2007].

Chapter 26: Stability in the Sense of Poisson

With regard to the concept of stability and the three-body problem, we can distinguish three features that we would like to investigate:

1. None of the three bodies can have unbounded motion.
2. No two of the bodies can collide, and the bodies have a positive minimum distance.
3. The system passes arbitrarily near to its initial conditions an infinite number of times.

If the third condition has been satisfied without knowledge about the first two, the system can be considered only *stable in the sense of Poisson*. For some models, the first and third conditions are satisfied, while about the second condition we know little.

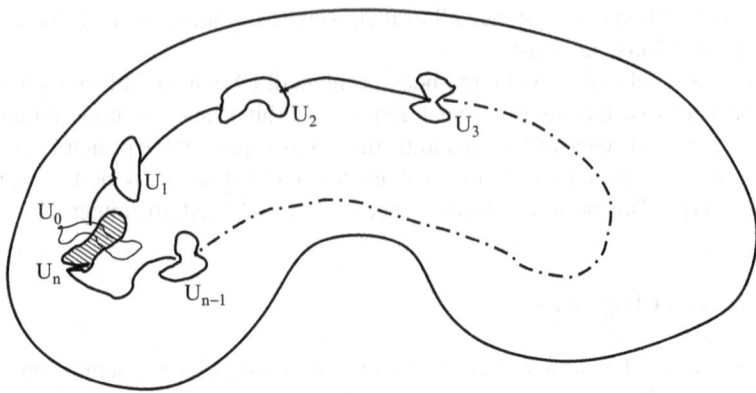

Fig. 9.8 An initial point P contained in a set U_0 with "volume" N_ε is, together with U_0, volume-or measure-preservingly mapped into a bounded domain. After repeating the map a finite number of times n, a number of points contained in U_0 will return to this set in the intersection of U_0 and U_n

The recurrence theorem is demonstrated first for an incompressible liquid contained in a nondeformable vessel. The velocity field of a particle is assumed to be autonomous, which gives a treatment that looks special but is in fact a general discussion of the three-dimensional case. Using the invariance of the volume of a liquid cell, it is proved that most particles return an infinite number of times arbitrarily near to their initial position. The same reasoning can be applied to negative time.

A set U_0 of particles with a positive volume will after a definite interval of time be changed into the set U_1, called the "consequent" of U_0. The preceding set, in this case U_0, is called the "antecedent" of U_1; see Figure 9.8.

In §296 it is noted that there exist particles that do not return to a neighbourhood of their initial position or do so only a finite number of times. A probability argument is given that the number of particles that return only a finite number of times is arbitrarily small.

In §297, the recurrence theorem is extended to autonomous systems of arbitrary finite dimension that have a positive invariant integral and have (in modern terms) a phase flow defined on a bounded domain.

One can apply the recurrence theorem to the restricted three-body problem of §9 (planar, two bodies in circular orbits with third body mass zero) with the conclusion that one has Poisson stability. The general three-body problem also admits the energy integral, but application of the recurrence theorem is not so simple. There are, for instance, solutions in which two of the bodies approach each other very closely while the third one moves far away.

Comments on Chapter 26

This chapter contains the famous recurrence theorem, originally formulated in the prize essay for Oscar II [Poincaré 1890b]. The probability argument for the statement that there is only a negligible number of orbits that do not return an infinite number of times is in the modern literature replaced by an equivalent measure-theoretic formulation: the set of points that are not recurrent has zero Lebesgue measure. The notion of measure came after Poincaré. In Figures 9.8 and 9.11, we have an area-preserving map that is recurrent. Instead of a volume-preserving flow in a bounded domain, we will consider a measure-preserving map of a bounded set into itself. For an illustration see Figure 9.8.

The recurrence time is not specified but will generally depend on the nearness to the initial position that one requires. Suppose we have an initial point P in a bounded set with "volume" V. An ε-neighbourhood of P has volume N_ε. We expect the return time to be roughly proportional to V/N_ε. This can be an optimistic or a pessimistic guess, depending on the dynamics and the set concerned. Also, we cannot expect the return time to be uniform over the set. For a discussion see [Ghys 2010].

In the first half of the twentieth century, the recurrence theorem led to heated discussions about its consequences for statistical mechanics. The second law of thermodynamics states that in a closed system, the entropy increases monotonically, producing a disordered dynamical state. A famous "paradox" involves two containers, one filled with a gas and the other one empty (vacuum). Connecting the containers will cause the gas to be evenly distributed over the two containers, but the recurrence theorem tells us that the gas will return to the first container after a finite time. The standard answer to this is that the time to spread the gas equally among the containers is very much shorter than the time scale of recurrence. However, this statement is more a quantitative prediction than a qualitative explanation, and it did not satisfy all scientists. See, for instance, [Steckline 1983] for the discussion between Ernst Zermelo (1871–1953) and Ludwig Boltzmann (1844–1906).

Émile Borel (1871–1956) discusses statistical mechanics and irreversibility in [Borel 1914, note II]. If the universe can be contained in a sphere with very large radius R, the recurrence theorem applies to the universe. However, this rests on two assumptions: first, that such a sphere exists, and second, that there is no long-time interaction with another universe at an extremely large distance. "Long time" is understood here as the time interval needed to observe the quasiperiodicity predicted by the recurrence theorem. According to Borel, such speculations are not within the realm of physics. A more probable state of the universe is to be expected as the outcome of evolution.

There is a curious link to Friedrich Nietzsche (1844–1900), who in 1880 developed his idea of the eternal recurrence of life. In his *Die Fröhliche Wissenschaft*, he writes on this eternal recurrence ("Die ewige Wiederkehr des Gleichen"):

> You will have to live this life one more time and then countlessly many times. (Dieses Leben wirst du noch einmal und noch unzählbare Male leben müssen.)

For the justification of these ideas, Nietzsche used to quote writers on mathematical physics who wrote about recurrence. Perhaps Poincaré would have been interested in discussing this, but no correspondence between the philosopher and the scientist is known.

A beautiful and instructive class of problems known as billiard dynamics was opened up by George D. Birkhoff (1884–1944) in [Birkhoff 1927]. The dynamics of the models is characterized by area-preserving maps and contains all the problems of periodic solutions, integrability, and chaos of low-dimensional Hamiltonian mechanics.

Chapter 27: The Theory of Consequents (Poincaré Map)

The notion of consequents is demonstrated first for a three-dimensional autonomous system with rotation around the z-axis. Considering a point M_0 in the coordinate plane $y = 0$, a solution starting at M_0 will return to a point M_1 of the plane $y = 0$, called a consequent of M_0. In the same way, because of continuity, a curve C_0 in the plane will be mapped by the ensemble of solutions into a curve C_1 in the plane; see Figure 9.9. If, in addition, we have an invariant integral that conserves volume, it is argued that the area encompassed by a closed curve in the plane is conserved for its consequents. In the case of the map of this closed curve, there are four possibilities:

1. C_1 is interior to C_0.
2. C_0 is interior to C_1.
3. The two curves are exterior to each other.
4. The two curves intersect.

If we have an integral that preserves area, the first two possibilities are excluded. If, in addition, the system has a small parameter μ and for $\mu = 0$, C_0 is an invariant curve, i.e., C_0 is mapped into itself, we will have for μ positive but small that C_1 will be a small deformation of C_0; it follows that in this case, we have also to exclude the third possibility: the curves have to intersect.

Consider a periodic solution transversal to the plane $y = 0$; starting at M_0 in the plane, its consequent will again be M_0. Nearby solutions can be parameterized by the initial conditions and form so-called *asymptotic surfaces*.

Consider now (still in our three-dimensional system) the case with a small parameter μ and assume that for $\mu = 0$, we have a closed invariant curve K. For $\mu > 0$, the consequent of K need not be closed, and five geometric possibilities for K and its consequent can be analysed; a few remain (§§308–309).

It is noted that the assumption of rotation around the z-axis and choosing the transversal plane $y = 0$ are not essential for the treatment of consequents. The implication is that we can apply the theory to the equations

$$\dot{x}_i = F_{y_i}, \quad \dot{y}_i = -F_{x_i}, \quad i = 1, 2,$$

Fig. 9.9 A point contained in a plane, transversal to the flow, is mapped again into this plane. A curve C_0 in the plane is mapped into a curve C_1. A fixed point of the map will correspond to a periodic solution

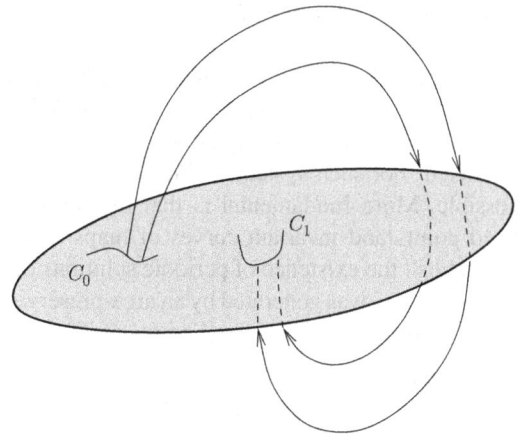

of dynamics with two degrees of freedom that we have studied before. Here F is periodic in y_1, y_2 and can be expanded with respect to the small parameter μ: $F = F_0(x) + \mu F_1 + \mu^2 F_2 + \cdots$, where F_0 depends on x_1, x_2 only. Fixing the energy $F = C$ and putting $\mu = 0$, we have

$$\dot{x}_i = 0, \quad \dot{y}_i = -\partial F_0/\partial x_i, \quad i = 1, 2.$$

So the x_i are constant and $y_i = n_i t + c_i$ (constant) with the n_i depending on the constant x_i. This means that for $\mu = 0$, the solutions are located on tori; the trajectories are closed if $n_1 : n_2$ is rational (commensurability); if the ratio is irrational, then we have quasiperiodic flow over the torus. Suppose that n_2 does not vanish and is positive for certain x-values. Then this will also hold for μ small. The plane $y_2 = 0$ can, for suitable x-values, be used to analyse the consequents of the solutions.

Consider now for $\mu = 0$ that a closed curve is generated by the intersection of a torus in the commensurable case. For $\mu = 0$, this is an invariant curve (of the consequents or Poincaré map), and we expect that for μ positive but small, certain periodic solutions survive by continuation. There will be at least two of them, with one stable and one unstable. Consider an unstable periodic solution and its finite number of intersections with the plane. Two asymptotic surfaces will pass through this unstable periodic solution. At each point of intersection of this periodic solution, the particular asymptotic surface passing through the periodic solution produces two curves. Analysis of the geometry of the plane of intersection shows that the third hypothesis of §308 is the only one remaining: two curves of consequents intersect. Actually, the demonstration is for period five and five intersection points, but this is not essential.

Application is again to the planar restricted three-body problem as formulated in §9. For instance, for small eccentricities, the reduction to the plane of intersection can be carried out. We obtain the intersection of the asymptotic curves.

Comments on Chapter 27

The introduction of Poincaré maps to study the dynamics of Hamiltonian systems
has been very influential. For systems with two degrees of freedom, one can,
by fixing the energy, study two-dimensional area-preserving maps of the plane
into itself. For such systems, which are four-dimensional, this makes visualization
possible. More fundamental is that this idea was the beginning of the study of
fixed points and invariant curves of maps. Fixed-point theorems have been used
to establish the existence of periodic solutions of dynamical systems.

Figure 9.11 was generated by an area-preserving map T_H. For instance, 6-periodic
solutions can be found as fixed points of T_H^6, a 10-periodic solution as a fixed point
of T_H^{10}.

Chapter 28: Periodic Solutions of the Second Kind (Superharmonics)

Consider an n-dimensional T-periodic system of the form $\dot{x} = X(x, t)$ with
T-periodic solution $\phi(t)$. We call this a solution of the first kind ("première genre").
Solutions near $\phi(t)$ with period kT, $k > 1$ a positive natural number, will be called
periodic solutions of the second kind ("deuxième genre") or superharmonic.

A few general observations can be made. Consider a neighbouring solution of
$\phi(t)$, starting at $\phi(0) + \beta$ (β small); at time $t = kT$, this solution will have the
value $\phi(kT) + \beta + \psi$; ψ will also be small and will be a function of β. In fact,
ψ will be expandable in powers of β. A necessary condition for this solution to be
kT-periodic is that $\psi = 0$ (§314).

We assume now in addition that the original n-dimensional differential equation
depends on a small parameter μ and that the period T does not depend on μ.
This means that our necessary condition depends on the $n + 1$ parameters β. The
necessary condition $\psi = 0$ corresponds to a curve in $(n+1)$-dimensional parameter
space. If $\beta = 0$, we have a straight line in parameter space corresponding to the
known T-periodic solution. Consider a point P of this straight line for which $\beta = 0$,
$\mu = \mu_0$. To have more than one branch corresponding to a periodic solution passing
through P, the Jacobian determinant $|\partial\psi/\partial\beta|$ has to vanish in P.

It is easy to show that a necessary condition for this is that one of the
characteristic exponents of the T-periodic solution of the original equation be a
multiple of $2\pi i/kT$ (§314).

The technical analysis in §315 is concerned with the bifurcational behaviour of
the solutions near P, assuming that not all the minors of the Jacobian at this point
vanish. A local series expansion enables us to decide whether periodic solutions of
the second kind exist. If all minors to a certain order vanish, there will be more than
one characteristic exponent that is a multiple of $2\pi i/kT$, and it is then possible that
more periodic solutions of the second kind exist.

To start with, we have considered an equation that is explicitly T-periodic. In the
case of an autonomous equation $\dot{x} = X(x)$ depending on a small parameter μ, we
will assume that there exists a T-periodic solution $\phi(t)$ that (of course) depends on

μ; if $\mu = 0$, the period is T_0. The modification is now that in repeating the analysis of §314, we will allow small time shifts for the multiple periods; instead of kT, we will consider $k(T + \tau)$, so that now we have $(n + 2)$ parameters. In parameter space, a surface will correspond to the T-periodic solution, and we will look for bifurcating sheets of this surface.

Another modification is necessary if the differential equation admits a first integral. In the case of a nonautonomous equation, this means that one characteristic exponent is zero, while in the autonomous case, two characteristic exponents vanish. This suggests a reduction of the dimension of parameter space by one in the nonautonomous case and two in the autonomous case.

In the case of the Hamiltonian equations of dynamics, an essential part is played by the function S defined as follows: F is the time-independent Hamiltonian; C a suitable constant; ξ, η the values of position and momentum x, y at $t = 0$; X, Y the values of x, y for $t = T$. Then

$$\frac{dS}{dT} = 2(F - C) - \sum \left[(X - \xi)\frac{\partial F}{\partial X} + (Y - \eta)\frac{\partial F}{\partial Y} \right].$$

The maxima and minima of S are candidates for the existence of T-periodic solutions. A large number of cases can be distinguished, among which are included a specification to two degrees of freedom and the case that there is more than one first integral. The analysis can be used to prove the existence of periodic solutions of the second kind with periods mT, $m = 2, 3, \ldots$.

Chapter 29: Forms of the Principle of Minimal Action

For the solutions of the equations of dynamics, the action J, given by

$$J = \int_{t_0}^{t_1} \left(-F + \sum y_i \frac{dx_i}{dt} \right) dt,$$

must be a minimum. Using canonical transformations, one obtains corresponding different forms of this minimum principle. Other, related, formulations of minimum principles are due to Maupertuis and to Hamilton. A necessary condition for the action integral J to be a minimum is that the first variation vanishes. We call this condition A. A second condition, B, will ascertain that we have actually a minimum; its formulation takes different forms for Hamilton's and Maupertuis's minimum principles. The principles of minimal action can be explicitly demonstrated for stable and unstable periodic solutions with a number of qualitative conclusions.

Chapter 30: The Formation of Solutions of the Second Kind

Consider the time-independent canonical equations of dynamics assuming two degrees of freedom and the existence of a periodic solution $(x, y) = (\phi(t), \psi(t))$. Introducing the localization of §274 involves a small parameter ε and the expansion of F in homogeneous polynomials of x and y. It is convenient to introduce two parameters, λ, μ, arising in the equations of motion, whereas for $\lambda = \mu = 0$, the equations admit the given periodic solution. We assume that λ, μ can be expanded in a power series with respect to ε.

By expanding the solutions and imposing the presence of solutions of the second kind (superharmonic), we can derive a consistent system of perturbation equations. This technical treatment is illustrated in §368 by the problem of §13, where $F = F_0 + \mu F_1 + \mu^2 F_2 + \cdots$ (μ plays the part of ε above). Here F_0 depends on x only. In §42 we established the existence of a periodic solution $(\phi(t), \psi(t))$ that can be developed in powers of μ. Suppose that its period is T and its two nonzero characteristic exponents are $\pm\alpha$; α depends on μ and can be developed in powers of $\sqrt{\mu}$. If αT is commensurable with $2\pi i$ for $\mu = \mu_0$, we conclude that a periodic solution of the second kind exists in a neighbourhood of $\mu = \mu_0$. To be more explicit, we consider in §369 the example of §199:

$$F = x_2 + x_1^2 + \mu \cos y_1.$$

Using an elliptic integral, we find that there exists a solution with period ω. In this case, starting with this exact solution, we obtain several periodic solutions of the first and second kinds by varying the initial conditions.

Chapter 31: Properties of Solutions of the Second Kind

This chapter focuses on systems with two degrees of freedom and one degree of freedom with a periodic forcing. In the first case, the positions (x_1, x_2) describe trajectories in a plane with various kinds of foci. Closed orbits correspond to periodic solutions characterized by two zero and two nonzero characteristic exponents (α) indicating stability or instability. A stable solution can exchange stability with another periodic solution at a critical value of the parameter α. Also, if a periodic solution changes stability, one can be certain that this is by exchange with another periodic solution. If two periodic solutions merge and vanish, than a pair of stable and a pair of unstable periodic solutions vanish.

As an illustration, one can consider a restricted three-body problem studied by G. H. Darwin using numerical techniques. Various periodic solutions can be identified, and several conjectures can be made regarding the connections between the families of periodic solutions (§§381–384).

Fig. 9.10 Transcritical bifurcation. The amplitude r is given as a function of the parameter μ. There are two solutions, $r = 0$ and $r = \mu$, which merge at $\mu = 0$. Stability is exchanged at this point as indicated by the arrows; a hashed line corresponds to an unstable solution, a solid line to a stable solution

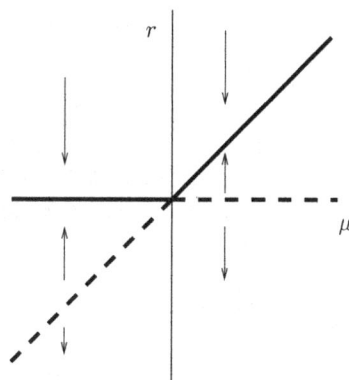

Comments on Chapters 28–31

These chapters are concerned with periodic solutions of the second kind with some observations on the variational approach in Chapter 29. Here we find for the first time the expression "Hamiltonian form." The chapters employ bifurcation methods to demonstrate the emergence and vanishing of periodic solutions in a setting that still looks modern. A simple example of exchange of stability is the case that the periodic solution is of the form $r_\mu \phi(t)$ with $\phi(t)$ T-periodic, with the amplitude r depending on the parameter μ. Suppose that the equation for r is given by

$$\dot{r} = \mu r - r^2.$$

In Figure 9.10, the behaviour of r as a function of μ is presented. Such bifurcational behaviour is called transcritical. It occurs often in applications.

Hopf bifurcation (see also Section 9.4), i.e., the emergence of a periodic solution when the characteristic exponents cross a certain value, is a natural part of the bifurcation analysis in Chapters 4 and 31.

Chapter 32: Periodic Solutions of the Second Kind

Consider again the equation of dynamics as in §§13 and 42. The energy function F can be expanded with respect to the small parameter μ as $F = F_0 + \mu F_1 + \mu^2 F_2 + \cdots$, $F_0 = F_0(x)$. It was shown that under certain conditions, especially if the Jacobian determinant $|\partial F_0 / \partial x|$ is nonzero, there exist T-periodic solutions; when the period is added to t, the variables $y_1, y_2, \ldots$ increase with $2k_1\pi, 2k_2\pi, \ldots$. If the Jacobian determinant vanishes, as it does in the three-body problem where $F_0 = F_0(x_1, x_2)$ with the number of degrees of freedom greater than two, we can still find periodic solutions for which $k_3 = k_4 = \cdots = 0$. Periodic solutions of the *second kind*, if they exist, are periodic solutions with $k_3, k_4, \ldots \neq 0$.

The construction runs as follows. If $\mu = 0$, the two smaller bodies (planets) move in Keplerian orbits; their masses are too small to influence each other. However, in the case of near-collision, called a "shock," the two bodies attract each other during a short time period, and their orbits change. After the near-collision, they move again in Keplerian orbits, but different ones. If the changes are not too large, the expansion procedure of §42 can be applied again to demonstrate the presence of periodic solutions of the second kind.

Comment on Chapter 32

This chapter is short but contains an interesting idea: the formation of periodic solutions when there are near-collisions (shocks). This possibility is mentioned already in Chapter 3. The chapter is more of a recipe for treating such cases than a detailed analysis. Analytically, this kind of problem can now be handled by singular perturbation techniques; numerically, near-collisions play a large part in the planning and control of space exploration involving the so-called slingshot phenomenon.

Chapter 33: Doubly Asymptotic Solutions

To illustrate the theory, we consider again the problem of §9, the restricted planar circular three-body problem. Assume that the masses of the main bodies are m_1, m_2 with $m_2 = 1 - \mu$, $m_1 = \mu$, where μ is a positive small parameter; the notation is different from that in Figure 9.2. The larger mass, m_2, will be put in the centre of the coordinate system at A, while the smaller mass, m_1 (point B), will move in a circular orbit with radius 1 around A; the mass C that is so small that it does not affect the motion of A and B has variable distance r_1 to B and r_2 to A.

With Hamiltonian $F = F_0 + \mu F_1$ and conjugate canonical variables x_1, x_2, y_1, y_2, we have that

$$F_0 = \frac{2}{(x_1 + x_2)^2} + \frac{x_2 - x_1}{2},$$

and $F_1(x_1, x_2, y_1, y_2)$ is periodic in the angles y_1, y_2.

We consider points in phase space with energy such that they have antecedents and consequents in a fixed transverse (Poincaré) plane P. As we have seen, for instance in §40, there exist periodic solutions of the first type. In some cases, the solutions are 2π-periodic with respect to y_1 and can be expanded in integral powers of μ.

A periodic solution produces in a transverse plane P a set of periodic points that we will call a "system of periodic points" or "periodic system." The unstable periodic solutions are associated with asymptotic surfaces (stable and unstable manifolds) that produce in the transverse plane P four curves emanating from each periodic point of a periodic system. Since there is an infinite number of unstable periodic solutions, there is an infinite number of periodic points with asymptotic curves in P.

Two stable asymptotic curves (characteristic exponents negative) of the same periodic system cannot intersect, but stable and unstable asymptotic curves can intersect at a point Q. The corresponding asymptotic surfaces intersect along a trajectory τ through Q; a remarkable solution σ is associated with τ. For $t \to +\infty$, the solution σ will approach a point, while for $t \to -\infty$, it will approach another point. This means that the solution σ is doubly asymptotic. In the case that the two limiting points coincide, the solution σ is called *homoclinic*, while if the points are different, σ is called *heteroclinic*.

The homoclinic solutions are analysed in §395. Considering the transverse plane P containing a periodic system, one can apply the energy integral to the map of P into itself to find that the map is area-preserving. Applying this to a polygon formed by asymptotic curves, one finds that there should exist at least two doubly asymptotic solutions. Further analysis shows that in fact, there exists an infinite number of them, corresponding to an infinite number of homoclinic solutions.

The next step in §396 is to show that if one finds a homoclinic point by the intersection of two asymptotic curves, there has to exist an infinite number of intersections of these asymptotic curves; in proving this, the area-preserving character of the map of P into itself again plays a part. Poincaré conjectures here that the doubly asymptotic solutions are everywhere dense on the asymptotic surface. One can consider two asymptotic curves in P with an infinite number of doubly asymptotic orbits entering and leaving a neighbourhood of these two curves. A famous description follows in §397:

> If one tries to represent the figure formed by these two curves with an infinite number of intersections whereby each one corresponds to a doubly asymptotic solution, these intersections form a kind of lattice-work, a tissue, a network of infinite closely packed meshes. Each of the two curves must not cut itself but it must fold onto itself in a very complex way to be able to cut an infinite number of times through each mesh of the network.
>
> One will be struck by the complexity of this picture, which I do not even dare to sketch. Nothing is more appropriate to give us an idea of the intricateness of the three-body problem and in general all problems of dynamics where one has not a uniform integral and where the Bohlin series are divergent.

The original text is as follows:

> Que l'on cherche à se représenter la figure formée par ces deux courbes et leurs intersections en nombre infini dont chacune correspond à une solution doublement asymptotique, ces intersections forment une sorte de trellis, de tissu, de réseau à mailles infiniment serrées; chacune des deux courbes ne doit jamais se recouper elle-même, mais elle doit se replier sur elle-même d'une manière très complexe pour venir recouper une infinité de fois toutes les mailles du réseau.
>
> On sera frappé de la complexité de cette figure, que je ne cherche même pas à tracer. Rien n'est plus propre à nous donner une idée de la complication du problème des trois corps et en général de tous les problèmes de Dynamique où il n'y a pas d'intégrale uniforme et où les series de Bohlin sont divergentes.

There are now various possibilities:

1. The set of intersections of asymptotic curves in the plane P corresponding to doubly asymptotic solutions fills up the whole plane. In such a case, we would conclude instability of the (solar) system.
2. The set of intersections of asymptotic curves fills up only a restricted part of the plane P. There can be stability or instability depending on the initial conditions.
3. The set of intersections of asymptotic curves is found in each interval but has area zero.

It is finally noted that in the case of doubly asymptotic solutions near a periodic solution, one can follow the antecedents and consequents of a point, finding that these points remove themselves from the periodic system and return to it. This will not be in a way that repeats itself, and moreover, the ordering of the projections of these points on the x- and y-axes will be irregular and different in the two dimensions (§398).

The information on heteroclinic solutions is more restricted (§391), but if there exists one heteroclinic solution, then there are infinitely many. The proof uses the geometry of the two-dimensional (Poincaré) map.

Proposition: If there exists one heteroclinic solution, then the Newcomb and Lindstedt series are nowhere convergent (§400).

It is possible to analyse examples of homoclinic solutions in §401 by considering the Hamiltonian

$$-F = p + q^2 - 2\mu \sin^2 \left(\frac{y}{2}\right) - \mu \varepsilon \sin y \cos x$$

with conjugate variables $(p, x; q, y)$ and small parameters μ, ε. One can find explicitly an equation for two doubly asymptotic solutions. There exist infinitely many, but to describe these, one has to compute higher-order approximations.

To have a more concrete example of heteroclinic behaviour, one looks in §403 at a system with

$$F = F_0(p, q, y) + \varepsilon F_1(p, q, x, y),$$

with F_0, F_1 periodic in x and y. Considering $F_0 = h$ (constant), p is a parameter, and q and y are coordinates in a plane. We are interested in double points corresponding to unstable periodic solutions. Such a point represents a doubly infinite set of such solutions, since h and p are free parameters. By expanding the Jacobi function S and Fourier analysing both S and F_1, we can derive explicit equations that may produce heteroclinic solutions.

An explicit example in §404 is

$$F_0 = -p - q^2 + 2\mu \sin^2 \left(\frac{y - y_0}{2}\right) \sin^2 \left(\frac{y - y_1}{2}\right),$$

$$F_1 = \mu \cos x \sin(y - y_0) \sin(y - y_1).$$

The heteroclinic solution will tend towards the periodic solutions given by

$$x = t, \quad p = q = 0, \quad y = y_0,$$
$$x = t, \quad p = q = 0, \quad y = y_1.$$

The two periodic solutions have equal angular frequencies $1, 0$, which makes the example rather degenerate.

In §405, a modification is considered with

$$F = (1 - \mu) F_0(x_1, x_2) + \mu F_1(x_1, y_1; x_2, y_2).$$

It is easy to give conditions such that for $\mu = 0$ ($F = F_0$), the system has two periodic solutions with respectively the angular frequencies $1, 0$ and $0, 1$. We define F_1 such that for $\mu = 1$ ($F = F_1$), the same two periodic solutions still exist. Using the method of Jacobi, we can obtain explicit expressions for the solutions in the case $\mu = 1$. Considering then a small perturbation of F_1,

$$F = F_1 + (1 - \mu)(F_0 - F_1),$$

we can show that for μ near 1, the asymptotic surfaces intersect and that heteroclinic solutions exist. The analysis is not complete, since we have no results for μ small.

Comments on Chapter 33

The final chapter introduces homoclinic and heteroclinic solutions that play a fundamental part in characterizing the nature of phase flow. The stable and unstable manifolds of an unstable periodic solution may intersect to produce one and consequently an infinite number of homoclinic solutions. The transversal flow that Poincaré "does not even dare to sketch" shows the chaotic character of the phase flow. This is typical for area-preserving maps that rule Hamiltonian systems with two degrees of freedom and in fact with any number of degrees of freedom. We show a picture of such a map in Figure 9.11.

This area-preserving map T_H has the following form:

$$\begin{pmatrix} x \\ y \end{pmatrix} \mapsto \begin{pmatrix} \cos\alpha & -\sin\alpha \\ \sin\alpha & \cos\alpha \end{pmatrix} \begin{pmatrix} x \\ y \end{pmatrix} + \sin x \begin{pmatrix} -\sin\alpha \\ \cos\alpha \end{pmatrix}. \qquad (9.6)$$

In Figure 9.11, we took $\alpha = 3\pi/5$. The closed KAM curves around the centre (discussed in the comments on Chapters 5 and 25) suggest that for small values of x and y, the map is nearly integrable. For larger values of x and y the chaotic nature of the map becomes more transparent. Of the possibilities Poincaré lists in Chapter 33, this would be an example of possibility 2: the intersection of asymptotic curves fills up only a restricted area of the transversal plane. It might be possible that actual

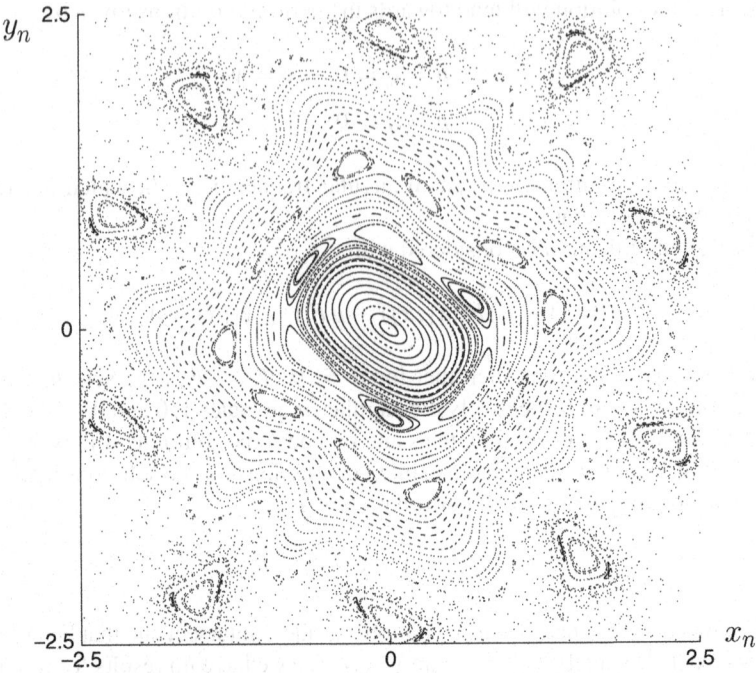

Fig. 9.11 The area-preserving map T_H produced by (9.6) for $\alpha = 3\pi/5$. In the centre of the plane there is a dominant family of closed KAM curves. In between the curves there are again stable and unstable periodic solutions, but they cannot be observed at this level of precision. Outside this family of closed curves one finds stable periodic solutions associated with unstable periodic solutions. Moving out, one observes a stable and an unstable 10-periodic solution and further on, another pair of 10-periodic solutions. The unstable solutions have stable and unstable manifolds that intersect an infinite number of times, producing chaotic behaviour. The dots correspond to orbits returning chaotically in the plane as the map is applied repeatedly. The folding process described by Poincaré can be seen dynamically only by considering a small square in the plane and following its subsequent mappings (figure courtesy of Igor Hoveijn)

configurations of the three-body problem are in such a situation, but this was proved only for a simple case that does not correspond to realistic initial values of the solar system.

The conjecture that there exists an infinite number of doubly asymptotic orbits has been confirmed by modern research. The dynamics is called "homoclinic chaos" and can be described by a so-called horseshoe map; see [Broer 2004, Arnold 1978, Arnold et al. 1988]; it is also worthwhile to consult the monograph [Siegel and Moser 1971]. An influential reference for more general dynamical systems is [Smale 1967].

9.4 Hopf Bifurcation and Self-Excitation

A bifurcation that plays a very prominent role in nonlinear dynamics is the Hopf bifurcation, also referred to as the Poincaré–Andronov–Hopf bifurcation. It may happen that an equilibrium point whose eigenvalues depend on parameters has two purely imaginary eigenvalues if one of the parameters (μ) assumes a critical value, say μ_0. In this case, depending on the nonlinearities, there may exist a nearby periodic solution. If the periodic solution emerges for $\mu < \mu_0$, it is called subcritical, while for $\mu > \mu_0$, it is supercritical.

The classical example is the Van der Pol equation

$$\ddot{x} + x = \mu\dot{x}\left(1 - x^2\right).$$

If $\mu = \mu_0 = 0$, the equilibrium $x = 0$, $\dot{x} = 0$ has two imaginary eigenvalues, and increasing μ produces a unique periodic solution around the origin in the phase plane.

For periodic solutions and fixed points of a map, there are analogous results, whereby one usually refers to generalized Hopf, Hopf–Hopf, or Neimark–Sacker bifurcation.

The first place where this bifurcation was formulated is [Poincaré 1892, §51]. Poincaré considers an equilibrium of an autonomous equation

$$\dot{x} = X(x)$$

in $\mathbb{R}^n$ and considers this equilibrium a periodic solution with arbitrary period. Suppose that there is a parameter μ in the equation and that $x_1 = x_2 = \cdots = x_n = 0$ is an equilibrium for any value of μ. We will look for a solution near the origin with initial value $x(0) = \beta$ (the lack of a subscript indicates the vector form) and $x(T) = \psi + \beta$. If we can determine T with $\psi = 0$ and nontrivial β, we have found a periodic solution.

Poincaré actually considers a neighbourhood of $\mu = 0$; this is implicit in his considerations. We suppose that $X(x)$ can be expanded in powers of the x_i and μ. It follows that ψ can also be expanded in powers of β and μ. Consider, as in [Poincaré 1892, §38], the Jacobian J, but now at $\mu = 0$, $x = 0$:

$$J = \left.\frac{\partial X}{\partial x}\right|_{\mu=0, x=0}.$$

Assume that the $n \times n$ matrix J has single eigenvalues; call them $S_1, S_2, \ldots, S_n$. Starting at $x(0) = \beta$, the variational solution near $x = 0$ in the sense of [Poincaré 1892] is

$$\beta e^{Jt}.$$

To first order in μ, the periodicity condition becomes

$$\beta e^{JT} = \beta + \psi.$$

The determinant of the Jacobian of ψ with respect to β will be equal to

$$\Delta = \left(e^{S_1 T} - 1\right)\left(e^{S_2 T} - 1\right)\cdots\left(e^{S_n T} - 1\right).$$

If $\Delta \neq 0$, we will have only the trivial solution $\beta = 0$, corresponding to the equilibrium solution $x = 0$. The condition $\Delta = 0$ to obtain a nontrivial solution corresponds to (at least) two eigenvalues S_i being purely imaginary and conjugate. This condition makes the existence of a small periodic solution branching off the equilibrium $x = 0$ possible, but we still have to consider the nonlinear terms to see whether a periodic solution actually emerges.

The eigenvalues S_i will depend on μ. Adding the condition that at the critical value we have $\mu = 0$, we have two conjugate imaginary eigenvalues $S_{i,j}$ with $dS_{i,j}/d\mu \neq 0$. We will call such a bifurcation value of μ a Hopf point. For a fairly complete discussion, see [Chicone 1999], and for an introduction, see [Verhulst 2000].

Poincaré considers in [Poincaré 1892, §52] as an example the equations formulated by Hill for the motion of the Moon: two second-order equations with one nontrivial equilibrium. The equilibrium corresponds to the Moon being in constant conjunction or opposition at constant distance from the Earth. The eigenvalues of the Jacobian as formulated above have two real values and two imaginary ones. The conclusion is that a periodic solution exists near this equilibrium in near opposition or near conjunction with an amplitude that grows with the small parameter $\sqrt{\mu}$. Since two eigenvalues are real, it will be unstable.

The classical example of the Van der Pol equation is easier to analyse. We rescale $\sqrt{\mu}x = y$ to obtain

$$\ddot{y} + y = \dot{y}\left(\mu - y^2\right).$$

The Jacobian J at $(y, \dot{y}) = (0, 0)$ and $\mu = 0$ has the eigenvalues

$$S_{1,2} = \frac{1}{2}\left(\mu \pm \sqrt{\mu^2 - 4}\right),$$

so for $\mu = 0$, we have a Hopf point. Applying the periodicity condition, we can find approximations for the amplitude that are proportional to $\sqrt{\mu}$ in y and a period that is near 2π; see [Verhulst 2000].

Poincaré's interest in wireless telegraphy induced him in [Poincaré 1908b] to use periodic solutions obtained by this type of bifurcation. Periodic solutions as produced by the Van der Pol equation are today called self-excited oscillations. In the case of wireless telegraphy, one has to study a magnetic field very far removed from its source. The question is then how to design an antenna as a source of radiation that enables us to direct the magnetic field in a suitable

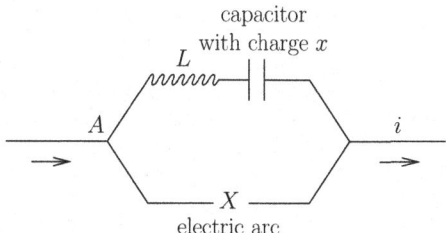

Fig. 9.12 The electrical circuit described in [Poincaré 1908b] to produce an oscillating magnetic field. At A, a constant current enters the circuit; L represents the self-induction, $1/H$ the capacitance of the capacitor, X the electrical arc, x the capacitor charge, i the outgoing current

way. Poincaré [Poincaré 1908b] goes into detail about various designs of the antenna. Although the wave is three-dimensional, because of the long distance, the component in the direction source–receiver is dominant. Neglecting damping effects, the radiation takes the form of a plane wave described by a Fourier integral (see also Chapter 11).

The magnetic field fluctuations are maintained by an electrical circuit as in Figure 9.12. The current in the branch with the capacitor is $x' = dx/dt$, while the current through the electric arc is $i + x'$. Considering the complete circuit, including the electric arc, the equation for the current becomes

$$Lx'' + \rho x' + \theta(x') + Hx = 0.$$

The constant ρ represents the resistance and other energy losses in the circuit, and $\theta(x')$ is the radiation term produced by the electric arc. Knowing a suitable function θ, one can construct isolated periodic solutions of this equation.

An example of such a function θ is the term suggested by Rayleigh in a different context:

$$\theta(x') = \mu x' \left(1 - x'^2\right).$$

It is not difficult to see that the equation containing this θ-function can be transformed into the Van der Pol equation (see [Verhulst 2000]).

9.5 The Poincaré–Birkhoff Theorem

In 1912, Poincaré submitted a theorem to the journal *Rendiconti del Circolo mathematica di Palermo* [Poincaré 1912b] without being able to present a proof. The reasons for this unusual step were given in the introduction:

> I have never made public a work that is so unfinished; so I believe it is necessary to explain in a few words the reasons that have induced me to publish it and to start with the reasons that brought me to undertake this. Already a long time ago, I have shown the existence of

periodic solutions of the three-body problem. However, the result is not quite satisfactory, for if the existence of each type of solution had been established for small values of the masses, one did not see what would happen for much larger values, which of the solutions would persist and in which order they would vanish. Thinking about this question, I became convinced that the answer would depend on a certain geometric theorem being correct or false, a theorem of which the formulation is very simple, at least in the case of the restricted problem and of dynamics problems that have not more than two degrees of freedom.

Poincaré adds that for two years he had tried to prove the theorem but without success. However, he was absolutely convinced that the theorem was correct. What to do? Let the matter rest?

> It seems that under these conditions, I would have to abstain from all publication of which I had not solved the problems. After all my fruitless efforts of long months, it seemed to me the wisest course to let the problem ripen and put it out of my mind for a few years. That would have been very good if I had been certain that I could return to it at some time, but at my age I could not say so. Also, the importance of the matter is too great and the quantity of results obtained already too considerable.

He had already some promising partial results and applications, and it seemed to be a waste to let all those ideas lie fallow. As it turned out, he was right. It is a beautiful fixed-point theorem that combines geometric thinking with dynamics. In [Poincaré 1916], it is classified under geometry, which is correct, but it also belongs under mechanics or differential equations. The theorem can be formulated as follows:

The Poincaré–Birkhoff Theorem

Theorem: *Consider in $\mathbb{R}^2$ the ring R bounded by the smooth closed curves C_a and C_b. The map $T : R \rightarrow R$ is continuous, one-to-one, and area-preserving. In applying T to R, the points of C_a move in the negative sense, and the points of C_b move in the positive sense (T is a "twist" map). Then T has at least two fixed points.*

The theorem was proved by Birkhoff [Birkhoff 1913] in a relatively simple way. It is difficult to understand why Poincaré did not produce such a proof. Looking at the 39 pages of [Poincaré 1912b], one has the feeling that Poincaré just for once saw too many small difficulties, that he got bogged down in details.

We now give an outline of Birkhoff's proof.

Proof of the Poincaré–Birkhoff Theorem [Birkhoff 1913]

Connect a point A of C_a with a point B of C_b by a straight line in R; see Figure 9.13. Since A is mapped in the negative direction, B in the positive direction, there must be a point on the straight line that moves in the radial direction only (continuity of T). Moving the straight line around the ring R, we find in this way a closed curve C of points in R that do not change their angle by the map T but move only in the radial direction. The curve TC produced by the map T applied to C is another

Fig. 9.13 Twist map T of a
ring-shaped domain R

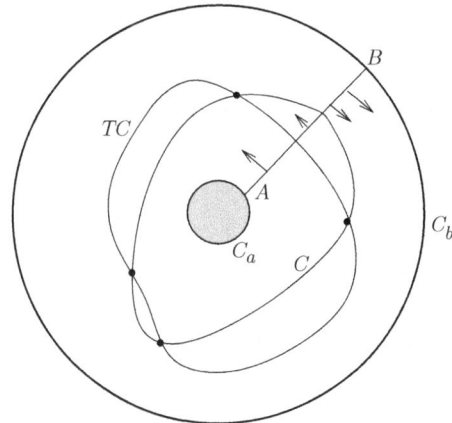

closed curve in R. Since the area between C_a and C is preserved under T, the curves
C and TC have to intersect at least twice. The points of intersection are fixed points
of T. □

The applications Poincaré had in mind can be indicated as follows. Consider a
dynamical system derived from a time-independent Hamiltonian with two degrees
of freedom. In studying the flow on a bounded energy manifold, one can make a
Poincaré section of the flow in a neighbourhood of a periodic solution. This periodic
solution is represented as a fixed point of the Poincaré map. If it is stable, the fixed
point will be surrounded by closed curves corresponding to invariant tori around the
periodic solution. The Poincaré map is area-preserving, so that the application of
the geometric theorem is possible if the twist condition has been satisfied. This can
be checked by considering the rotational properties of the map on the closed curves.

An interesting aspect is that if one is able to apply the theorem to a Hamiltonian
system, one finds not only two fixed points corresponding to two periodic solutions,
but an infinite number. This is caused by the presence of an infinite number of the
tori, enabling us to construct an infinite number of rings. If the tori are close, the
twist will usually be "small," and in this case, the period of the periodic solutions
will be large.

Other applications can be found in problems of three-dimensional divergence-
free flow and conservative billiard dynamics.

Chapter 10
Analysis Situs

In the middle of the nineteenth century, Michel Chasles strongly advocated that geometry and analysis be considered complementary disciplines, not to be separated if one wanted a complete picture of a mathematical theory. To put it simply, analysis provides shortcuts and routine in proofs, while geometry gives insight, showing the meaning of the results. Henri Poincaré's dissertation advisor, Gaston Darboux, was a student of Chasles. Poincaré was 12 years younger than Darboux, but he underwent a similar influence by studying the writings of Chasles. It shows in his early treatment of ODEs, where he introduced the geometry of the flow near critical points (equilibria) and used projection methods to clarify the structure of solution space. His geometric ideas helped him to handle automorphic functions, where he proposed the relationship between singularities of linear differential equations, Riemann sheets, and non-Euclidean geometry. That influence also came out abundantly in his analysis of dynamical systems, including conservative systems. His concept of "consequents" (Poincaré map) led him to formulate fixed-point theorems to obtain periodic solutions; see Section 9.5. The dynamics of high-dimensional dynamical systems with homoclinic and heteroclinic solutions together with their doubly asymptotic manifolds required subtle analysis in combination with geometric visualization. Also, in his work on the Laplace and Poisson equations, Poincaré's balayage method clearly pictures the analytic tool of shifting mass distributions in a convenient way.

So it is not surprising that Poincaré also considered the structure and characteristics of manifolds and other geometric objects without a context of applications. While doing this, he developed a whole new mathematical discipline, "analysis situs," or "topology." The term "analysis situs" can be traced to Leibniz, who was interested in what is common in the many geometric structures that one can imagine.

10.1   Early Topology

The first major result in topology stems from Euler. It is called the *Euler character-istic* or *Euler polyhedron formula*. Consider a convex two-dimensional polyhedron in $\mathbb{R}^3$ with V the number of vertices, E the number of edges, F the number of faces. The Euler characteristic χ is an invariant of such polyhedra:

$$\chi = V - E + F = 2.$$

The concept can be extended to more general closed surfaces. In this way, the Euler characteristic χ of the sphere in $\mathbb{R}^3$ is equal to 2; for the torus it is 0.

Another way to classify closed surfaces such as a sphere or a torus is by determining the number of its holes or handles. After Abel, this number is called its genus g, so for the sphere S^2 in 3-space, $g = 0$; a sphere with k handles has $g = k$. for the torus T^2, $g = 1$. The genus g of a closed surface in $\mathbb{R}^3$ is related to its Euler characteristic by the formula

$$\chi = 2 - 2g.$$

However, it is difficult to give a rigorous mathematical definition or characterization of a hole in a geometric object. This is one of the topics discussed in Poincaré's papers on analysis situs.

Before Poincaré, it was Enrico Betti (1823–1892) who generalized some of these notions to arbitrary dimensions. To characterize surfaces, Betti introduced the numbers $P_1, P_2, \ldots$, called Betti numbers after him. He was inspired by the ideas of Riemann on the connectivity of surfaces. Consider a closed surface in $\mathbb{R}^3$. The surface is connected if any two points on the surface can be connected by a curve that is entirely contained in the surface. Both a sphere and a torus are connected, so connectivity does not distinguish them. Imagine now a closed curve on a sphere. Any closed curve separates the sphere into two parts with points that cannot be connected without crossing the closed curve. The maximum number of closed curves that *do not* separate the sphere is zero, and this is just the Betti number P_1; for the sphere, $P_1 = 0$. For a torus, we can trace a closed curve around the hole or through the hole without separating the torus into two parts that have no connection, and so for the torus, $P_1 = 1$. See Figure 10.1. The concept of Betti number P_1 coincides with genus. In general, the Betti number P_1 of the closed surface S describes the maximum number of closed disjoint curves on S that do not separate the surface in 3-space.

In addition, Betti introduced P_2 for three-dimensional manifolds in $\mathbb{R}^4$, P_m for $(m + 1)$-dimensional manifolds in $\mathbb{R}^{m+2}$. In this framework, Betti added the concept of boundary to define this number. Think again of a two-dimensional surface S in $\mathbb{R}^3$. Instead of thinking of a separating closed curve, one can imagine a closed curve that forms the boundary of a two-dimensional connected submanifold of S. The maximum number of disjoint curves that fail to form such a boundary equals

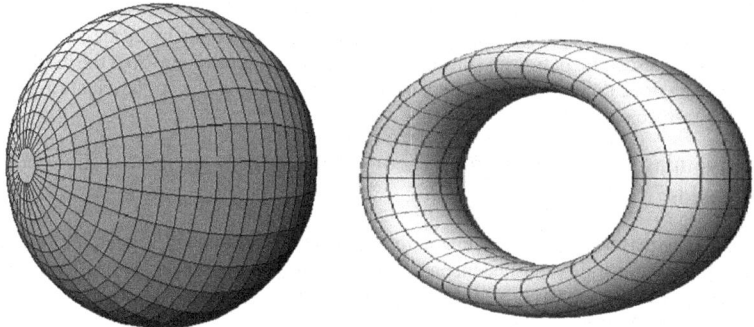

Fig. 10.1 (Left) A sphere (genus 0). A closed curve on the sphere separates it into two parts. (Right) A torus (genus 1). A closed curve through the hole in the torus or a closed curve encircling the hole does not separate it

the Betti number. For the 2-torus we have one closed curve through the hole and one closed curve encircling the hole.

At this point, Poincaré came up with completely new ideas in his analysis situs. These ideas and the development of topology after Poincaré are very technical, so in this section we present only a sketch of the theory, while referring for details to [Poincaré 2010, Novikov 2004] and their references; see also the collected works [Poincaré 1916, Vol. 6].

10.2   The Analysis Situs Papers

One of the first steps was a generalization of the Euler polyhedron formula. Consider a polyhedron in $\mathbb{R}^n$. Instead of vertices, edges, and faces, we associate with the polyhedron similar objects with the numbers $N_0, N_1, \ldots, N_{n-1}$, where the index indicates the dimension of the object. The generalization of Euler's formula for the polyhedron is then

$$N_0 - N_1 + N_2 - \cdots = 1 - (-1)^n.$$

For $n = 3$, we again have Euler's formula.

Poincaré's papers on topology were published in the period 1892–1905, starting in 1892 with a *Comptes Rendus* note, published by the Académie des Sciences. The monograph [Poincaré 2010] contains an introduction and translations into English. The papers were published in various journals and were written in discourse style, developing the subject and elucidating ideas in later supplements while correcting mistakes. One should read the papers with this in mind.

The first big paper in 1895 occupies 121 pages in the *Journal de l'École Polytechnique*. The introduction outlines its philosophy (translation by John Stillwell in [Poincaré 2010]):

> Nobody doubts nowadays that the geometry of n dimensions is a real object. Figures in
> hyperspace are as susceptible to precise definition as those in ordinary space, and even if
> we cannot represent them, we can still conceive of them and study them. So if the mechanics
> of more than three dimensions is to be condemned as lacking in object, the same cannot be
> said of hypergeometry.

After explaining the need for group theory and a new language for this geometry,
Poincaré adds:

> Perhaps these reasons are not sufficient in themselves? It is not enough, in fact, for a science
> to be legitimate; its utility must be incontestable. So many objects demand our attention that
> only the most important have the right to be considered.

Consider as an example the plane $\mathbb{C}$ and the group Γ generated by two Euclidean
translations in two different directions. The group Γ has a *fundamental domain* D,
a parallelogram, all translates of which fill the plane $\mathbb{C}$. In other words, the plane $\mathbb{C}$
is filled by the translates γD for $\gamma \in \Gamma$.

In this example one can introduce a calculus on the quotient manifold $\mathbb{C}/\Gamma$,
which is a torus. This quotient arises by identifying the opposite sides of the
fundamental parallelogram to produce a torus with two closed curves. In this
example, Poincaré's inspiration came from Fuchsian functions, the group Γ coming
from elliptic (periodic) functions and probably also from Hamiltonian mechanics,
where in the integrable cases, tori abound. A natural step is then to consider a group
of curves on the torus isomorphic to the group of translations of the plane $\mathbb{C}$. In this
example, this is the *fundamental group*. More generally, we can study topological
objects by identifying fundamental groups and algebraic quotient operations.

In the language of the 1895 analysis situs paper, this would be expressed thus:
Consider a manifold V, and let

$$F_1, \quad F_2, \quad \ldots, \quad F_\lambda,$$

be λ functions of the n coordinates of a point M of the manifold. One can consider
the values assumed by the functions $F_1, \ldots, F_\lambda$ when they begin in M and describe
a closed contour on V. The substitutions (as Poincaré calls them) undergone by the
functions F as the point M describes all possible closed contours that can be traced
on V form a group, called g.

The contours $v_1, v_2, \ldots$ are submanifolds of V, and we can follow the substitutions generated by them by adding and subtracting. If q-dimensional W is a
submanifold of p-dimensional V, then the boundary of W consists of $(q-1)$-
dimensional manifolds $v_1, v_2, \ldots$. An expression like

$$k_1 v_1 + k_2 v_2 \sim k_3 v_3 + k_4 v_4$$

means that the boundary of W is composed of k_1 manifolds similar to v_1, k_2
manifolds similar to v_2, k_3 manifolds similar to v_3 but oppositely oriented, and
k_4 manifolds similar to v_4 but oppositely oriented. Such relations are called
homologies. They play a part in characterizing the fundamental group.

Homology theory became one of the central tools for defining Betti numbers and their connection to the Euler characteristic. Consider as a simple example a disk, i.e., the set of all points in a plane bounded by a circle. A circle (submanifold) contained in this disk is homologous to a point. But if we consider as another simple example a two-dimensional annulus, then a circle in the annulus that is concentric with the bounding circles is not homologous to a point. In a similar way, we consider q-dimensional submanifolds W of a p-dimensional manifold V with their respective homologies.

As in analytical geometry, Poincaré began by describing manifolds through algebraic equations. A new constructive step was to represent a manifold by a collection of simplices. A one-dimensional simplex is a straight line segment, a two-dimensional simplex is a triangle, a three-dimensional simplex is a tetrahedron, etc. The sides of a polyhedron can be split up in triangles. How do we represent a smooth manifold by simplices? As an example, think of a sphere covered by many small triangles. Circles on the sphere are replaced by closed polygonal lines with one-dimensional segments for simplices. In this way, we can study homology by triangulating a manifold, producing a so-called *simplicial complex*. The geometrical meaning of cycles, submanifolds, and their homologies is preserved on such polyhedra. Triangulation of manifolds and the corresponding polyhedral structures are then used to develop an algebraic theory of topology.

With the concepts of fundamental group, homology, and other newly formulated concepts such as torsion coefficients, topology proceeded to develop as an independent subject. Needless to say, Poincaré left many problems unsolved; see again [Poincaré 2010].

10.3 The Poincaré Conjecture

The deepest of these problems left unsolved is the so-called *Poincaré conjecture*. It is typical that at first, Poincaré thought it so obvious that it needed no proof. And so it seems. The conjecture is formulated for n-dimensional manifolds, but the most difficult case turned out to be $n = 3$. Loosely speaking, in this case, the conjecture asserts that every simply connected closed three-dimensional manifold in $\mathbb{R}^4$ is homeomorphic to a 3-sphere. In other words, a closed surface without holes can be deformed continuously to produce a sphere. Trivial? Or is it perhaps not so trivial after all?

In his supplements on analysis situs, Poincaré presents preliminary versions of the conjecture while developing an ever more powerful armamentarium of techniques. At the end of the fifth supplement, in 1904, he finally formulates the conjecture as one of the questions that still have to be dealt with. In his own words [Poincaré 1916, Vol. 6]:

Considérons maintenant une variété V á trois dimensions. ... Est-il possible que le groupe fondamental de V se réduise à la substitution identique, et que pourtant V ne soit pas simplement connexe?

(Consider now a three-dimensional manifold V. ... Is it possible that the fundamental group of V can be reduced to the identity substitution yet V is not simply connected?)

For two-dimensional closed surfaces, there was soon a simple classification: a closed two-dimensional surface can be deformed continuously into a two-sphere if and only if it is simply connected. It is clear how technically complicated the problem is from the fact that the first proof in this field for higher dimensions was not for 3-manifolds but for n-manifolds with $n \geq 5$. It was found by Stephen Smale in 1961. A proof for 4-manifolds was found by Michael Freedman in 1982. The proof of the original Poincaré conjecture for 3-manifolds was given by Grigori Perelman in 2002–2003. The extensive research devoted to proving the conjecture by many mathematicians has led to a focus on the study of manifolds and thereby to a wealth of interesting new insights and knowledge. During the past century, the general theory of manifolds has grown enormously, and many interesting examples of 3-manifolds were discovered.

Chapter 11
Mathematical Physics

In this chapter, we will first look at new methods developed for partial differential equations, and in the following subsections, at a number of applications and physical theories. We aim at conveying the ideas while leaving technical details to the literature cited. We will leave out dynamical systems, since they were discussed in a separate chapter.

The sections on rotating fluid masses and cosmogony are based on the lecture notes [Poincaré 1890a]. For other sections, we use several papers, especially from Volumes 9 and 10 of the collected works.

11.1 Partial Differential Equations

Most scientists know about the work of Poincaré from a certain angle, such as celestial mechanics, topology, or automorphic functions. Those with such a one-sided view may be surprised to learn how many other topics he studied. An example is the theory of partial differential equations, which is basic to mathematical physics, for instance the theory of wave and heat propagation and potential theory. The theory and applications of partial differential equations form a very broad subject, and we will give only a sketch of the results obtained by Poincaré, describing the provenance and context of the ideas. The survey [Mawhin 2010] gives additional details and ends with the following conclusion:

> Poincaré's contributions to the equations of mathematical physics would have sufficed to place him among the greatest mathematicians of the end of the 19th and the beginning of the 20th century.

In 1890, Poincaré [Poincaré 1890c] noted the great similarities among the equations from very different fields of physics and chemistry. Considering the static or dynamic theory of electricity, optics, the theory of heat, elasticity, or hydrodynamics, one is always led to the study of the same group of differential equations, with

F. Verhulst, *Henri Poincaré: Impatient Genius*, DOI 10.1007/978-1-4614-2407-9_11,
© Springer Science+Business Media New York 2012

as primary example the Laplace equation. In addition, the boundary conditions that supplement the equations exhibit this similarity. Clearly, then, one should pay special mathematical attention to these typical equations.

The Balayage or Sweeping Method

In considering the Newtonian attraction properties of bodies with a given distribution of mass, one is led to a study of the Laplace and Poisson equations; see, for instance, [Poincaré 1890a, nr. 10]. One of the basic problems is to solve the equation

$$\Delta V = \frac{\partial^2 V}{\partial x^2} + \frac{\partial^2 V}{\partial y^2} + \frac{\partial^2 V}{\partial z^2} = 0 \quad \text{in} \quad D,$$

with $D \in \mathbb{R}^3$ a bounded domain. A twice differentiable function satisfying the Laplace equation $\Delta V = 0$ is called *harmonic*. If we require that on the boundary S of D we have $V = \Phi$ with Φ a known function, then this is called the Dirichlet boundary value problem for the Laplace equation. If we looked for solutions on the infinite domain exterior to D with the same boundary condition and certain conditions at infinity, we would have the exterior Dirichlet problem for the Laplace equation; this describes the Newtonian gravitational field in the empty space outside a body filling up D; the boundary potential prescribed on S derives from the interior distribution of mass. The exterior Dirichlet problem describes at the same time the electrical force field *outside* a conductor in D with prescribed electrical charge on S.

An elegant method for solving the Dirichlet boundary value problem is to consider the functional

$$I(V) = \int_D \|\nabla V\|^2 \, dx \, dy \, dz,$$

with V an element of the set of twice differentiable functions on D that are continuous on $D \cup S$. The nabla operator ∇ produces the gradient of V, and $\|\cdot\|$ is the Euclidean norm. The Dirichlet principle states that if one minimizes the functional $I(V)$ over the subset of functions V that satisfy the boundary condition, this solves the boundary value problem.

Around 1890, the validity of the Dirichlet principle was not yet proved, and indeed, Weierstrass had cast doubt on it, so scientists were looking for alternative solution methods.

Poincaré's balayage or sweeping method was published in [Poincaré 1890c]; a didactic presentation is given in [Poincaré 1890a, nr. 10]. In the lecture notes, the balayage method is explained for the interior of a sphere, after which more complicated geometries can be studied.

Consider a ball B with spherical surface S, centre O, and radius a in three-dimensional space. A point M is located inside the sphere at distance ρ from O.

Fig. 11.1 Balayage of a
spherical body

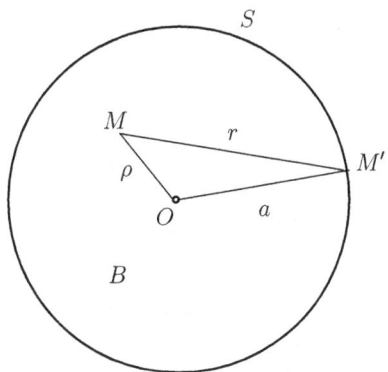

A surface element $d\omega'$ has centre of gravity M', and the distance from M to M' is
r; see Figure 11.1.

Inside and outside the sphere, we have a distribution of mass; at the centre of
gravity of a volume element $d\tau$, the density is μ. The gravitational potential V at a
point is written as

$$V = V_1 + V_2,$$

with V_1 the potential due to the mass interior to the sphere, V_2 the potential due to
the mass exterior to the sphere. As we know from potential theory, the potential V_1
will not change when we replace each mass $\mu\, d\tau$ in the ball B by a mass layer on
the surface S with density in M':

$$\mu' = \frac{a^2 - \rho^2}{4\pi a r^3} \mu\, d\tau.$$

Note that in replacing a volume by a surface element, we have an incorrect
presentation of the physical units; this will happen often in this treatment. The mass
distribution on the surface S is called the *equivalent layer*. With this procedure, we
perform a sweeping (balayage) of all the mass in the interior, producing in M' on S
the density

$$\mu'' = \int_B \mu \frac{a^2 - \rho^2}{4\pi a r^3}\, d\tau.$$

The integration is taken over the interior of the sphere. Clearly, $\mu'' > 0$. We put

$$U_1 = \int_S \frac{\mu''}{r}\, d\omega'.$$

The potential U_1 equals V_1 outside S, and we have $U_1 \leq V_1$ inside S. The potential
U_1 is called *subharmonic*.

Consider now, more generally, a bounded connected domain T with smooth
boundary surface S; see Figure 11.2. In the figure, P is a point (inside) T where a

Fig. 11.2 Balayage of a
bounded connected domain T

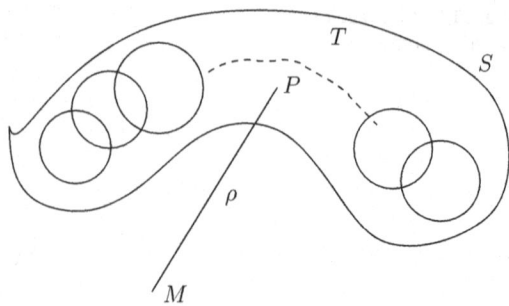

mass $m = 1$ is located. A point M in space at distance ρ undergoes a Newtonian
attractive force from m determined by the potential

$$V = \frac{m}{\rho}.$$

There exists a Green's function G with respect to the domain T with the
properties

$$G = \frac{1}{\rho} - H,$$

where H is harmonic and determined by

$$\Delta H = 0 \text{ in } T \quad \text{and} \quad H = \frac{1}{\rho} \text{ on } S.$$

In practice, the actual construction of Green's functions is restricted to simple
geometries. Consider now a function V that coincides with H in the domain T and
equals $1/\rho$ outside the boundary S; V is harmonic in T and outside S, continuous
in S, and regular at infinity. For a surface element $d\omega'$ as formulated above, the
theory of Green's functions produces the potential

$$V = -\frac{1}{4\pi} \int_S \frac{\partial G}{\partial n} \frac{d\omega'}{r},$$

so we can consider V the potential of a material layer spread out over S with density

$$-\frac{1}{4\pi} \frac{\partial G}{\partial n}.$$

For a point outside S, the potential from a mass in P equals the potential of the
layer on S.

To extend the balayage of a sphere to the domain T, we use a covering of T
by a denumerable set of balls and a corresponding sequence of harmonic functions.
In this sequence, each potential U_{n+1} is obtained from the preceding one U_n by

balayage. It requires a subtle argument, for which we refer to [Poincaré 1890a, nr. 10] and [Mawhin 2010].

In modern potential theory, one considers domains in $\mathbb{R}^n$ with Borel measures (instead of mass distributions) on sets of a general nature where the balayage produces another suitable measure.

Spectral Analysis: Estimates of Eigenvalues

Consider the heat (or diffusion) equation

$$\frac{\partial \phi}{\partial t} = \Delta \phi$$

on a bounded domain $D \in \mathbb{R}^3$ with smooth boundary S and a boundary condition $\phi|_S, t \geq 0$. Initially, when $t = 0$, we prescribe $\phi(x, y, z, 0)$. In looking for solutions in a form that separates space and time, we put

$$\phi(x, y, z, t) = U(x, y, z)T(t).$$

For U, this leads to the following equation on D with spatial derivatives only:

$$\Delta U + kU = 0.$$

The boundary condition, if it is linear, also separates; we choose the mixed condition

$$\frac{\partial U}{\partial n} + hU = 0 \quad \text{on } S,$$

where $\partial U / \partial n$ is the exterior normal derivative and h is a given constant. For the real number k, we will find a denumerable set of values such that U satisfies the boundary condition. We indicate these solutions—the eigenfunctions—by U_k. The corresponding numbers $k_1, k_2, \ldots$ are called the eigenvalues. One usually normalizes

$$\int_D U_k^2 \, dx \, dy \, dz = 1.$$

Any linear combination of eigenfunctions will solve the spatial boundary value problem. The separation process is called separation of variables or Fourier analysis. In Poincaré's time, the spectral analysis of such eigenvalue problems was not rigorous except for certain simple geometries of D.

A first step is the estimation of the eigenvalues by variational techniques. In the period 1887–1890, Poincaré formulated an extension of the Dirichlet principle; see [Poincaré 1890c]. Consider the expression

$$B(F) = h \int_S F^2 d\sigma + \int_D \|\nabla F\|^2 \, dx \, dy \, dz$$

and the normalization condition

$$A(F) = \int_D F^2 dx \, dy \, dz = 1.$$

The method of Lagrange multipliers gives us that the first eigenfunction U_1 with corresponding eigenvalue k_1 minimizes $B(F)$ over the set of nontrivial C^1 functions that satisfy $A(F) = 1$. Moreover, we have

$$k_1 \leq \frac{B(F)}{A(F)}$$

for all nontrivial functions F. Minimizing $B(F)$ over the smaller set of functions satisfying

$$\int_D F^2 \, dx \, dy \, dz = 1, \quad \int_D F U_1 \, dx \, dy \, dz = 0,$$

we obtain the second eigenfunction U_2 with eigenvalue k_2. This minimization process can be continued. In addition, Poincaré also obtained an upper bound for the eigenvalues. Introducing F as a linear combination

$$F = \sum_{j=1}^{n} \alpha_j F_j$$

and introducing the set S_n over which minimization will be carried out,

$$S_n = \left\{ \sum_{j=1}^{n} \alpha_j F_j : \alpha \in \mathbb{R}^n \right\},$$

it follows that we have the minimax characterization of eigenvalues

$$k_n = \min_{S_n} \max_{F \in S_n} \frac{B(F)}{A(F)}.$$

The last result is implicit in Poincaré's calculations. By considering lower bounds for the eigenvalues, Poincaré could show that $k_j \to \infty$ as $j \to \infty$.

Spectral Analysis: The Poincaré Inequality

Consider a convex open set $D \in \mathbb{R}^3$ with smooth boundary S. We are looking again for the solutions of the eigenvalue problem

$$\Delta U + kU = 0 \text{ in } D, \quad \frac{\partial U}{\partial n} + hU = 0 \text{ on } S.$$

In a long paper published in 1894 [Poincaré 1894b], Poincaré returned to the question of estimating the eigenvalues. By manipulating the integrals of the functionals derived earlier, he formulated an inequality for the second eigenvalue,

$$k_2 \geq \frac{6K_0\mu}{\pi d^5},$$

with μ and d respectively the volume and the diameter of D, and K_0 a numerical constant. This estimate can be generalized for an arbitrary eigenvalue k_n and also for a more general domain, as long as it can be decomposed into a finite number of convex subdomains. The estimates involve long calculations in polar coordinates.

An important generalization by Poincaré is to give bounds on a function in terms of its derivatives and the geometry of the domain of definition; such bounds are today formulated in norms corresponding to so-called Sobolev spaces. A typical result for a continuously differentiable function V defined on a convex set D in three-dimensional space such that $\int_D V \, dx \, dy \, dz = 0$ is

$$\int_D V^2 \, dx \, dy \, dz \leq c \int_D \|\nabla V\|^2 \, dx \, dy \, dz,$$

with suitable, but at this stage unknown, constant c.

In studies appearing after Poincaré, the estimates were extended for eigenvalue problems in $\mathbb{R}^n$, and optimal values for the numerical constants were obtained.

Rigorous Spectral Analysis

The proof of the existence of an infinite number of eigenvalues and eigenfunctions for the Dirichlet problem is usually attributed to Fredholm and Hilbert. The first proof, however, was given by Poincaré in [Poincaré 1894a, Poincaré 1894b]. Consider again the heat equation $\partial \phi / \partial t = \Delta \phi$ on a bounded domain $D \in \mathbb{R}^3$ with smooth boundary S and boundary condition $\partial U / \partial n + hU = 0$; h represents the emission coefficient of heat from the surface. Assuming that the body contains a source of heat, the equation is modified to

$$\frac{\partial \phi}{\partial t} = \Delta \phi + q,$$

where we have to specify q as a function of space and time. With simplifying assumptions, the separated equation becomes

$$\Delta U + \xi U + f = 0 \text{ in } D, \quad \frac{\partial U}{\partial n} + hU = 0 \text{ on } S,$$

with f a constant, ξ a parameter. Following the approach of Schwarz, Poincaré expands the solution in powers of ξ:

$$U = U_0 + \xi U_1 + \xi^2 U_2 + \cdots .$$

For the coefficients, one obtains the sequence of equations

$$\Delta U_0 + f = 0, \quad \Delta U_n + U_{n-1} = 0, \quad n = 1, 2, \ldots .$$

The boundary conditions for the U_n are inherited from the boundary condition for U. Considering first the Dirichlet problem $U = 0$ on S, one can construct the U_n using the Green's function for the Laplace operator. From these integral expressions, one can estimate the U_n to conclude that the series converges absolutely and uniformly with positive radius of convergence in ξ. In the construction, functions P_k arise that are called harmonic; they satisfy the equation

$$\Delta P_k + k P_k = 0.$$

The number k is called a characteristic value. Today we call P_k an eigenfunction and k the corresponding eigenvalue.

Generalized Solutions

Chaper VI ("Inégalités diverses") in [Poincaré 1894b] is concerned with maximum principles. Mawhin [Mawhin 2010] observes that here, for the first time, the concept of a generalized function is formulated. The reasoning goes as follows. Consider again a bounded domain D with smooth boundary S; we are looking for solutions of

$$\Delta u + f = 0 \quad \text{in } D,$$

with on S the boundary condition $\partial U / \partial n + hU = \phi$, where ϕ is a given function. The function v is continuous in the domain D, has continuous derivatives, and is otherwise arbitrary. Green's integral theorem produces, for the functions u and v,

$$\int_S \left(v \frac{\partial u}{\partial n} - u \frac{\partial v}{\partial n} \right) d\omega = \int_D (v \Delta u - u \Delta v) \, d\tau.$$

If u satisfies the equation and the boundary condition, we derive from this

$$\int_D vf\, d\tau + \int_D u\Delta v\, d\tau + \int_S v\phi\, d\omega = \int_S u\left(hv + \frac{\partial v}{\partial n}\right) d\omega.$$

Poincaré calls this a "modified condition" derived from the boundary value problem. If the modified condition is satisfied for arbitrary functions v, we can consider u a solution of the boundary value problem as long as u and $\partial u/\partial n$ exist and are continuous [Poincaré 1894b, p. 156].

Convergence in the Mean for Cooling Problems

In [Poincaré 1890a, nr. 8], from lectures given in 1893–1894, the classical convergence of series expansions for nonstationary heat flow was based on methods devised by Cauchy. In [Poincaré 1894a, Chapter 3], Poincaré notes the following problem: Consider again the boundary value problem for a cooling body in the form

$$\frac{\partial V}{\partial t} = \Delta V \text{ in } D, \quad \frac{\partial V}{\partial n} + hV = 0 \quad \text{on } S.$$

At the initial time, say $t = 0$, the temperature $V = V_0(x, y, z)$ is given, but as we will show, this may present a problem. One can derive the derivatives at $t = 0$, for instance

$$\frac{\partial V}{\partial t} = \Delta V_0, \quad \frac{\partial^2 V}{\partial t^2} = \Delta^2 V_0,$$

leading to an expansion of the form

$$V = V_0 + t\Delta V_0 + \frac{1}{2}t^2\Delta^2 V_0 + \cdots.$$

It is strange that the shape of the domain D does not enter into this expansion. To be more explicit, consider a one-dimensional domain $-\pi \leq x \leq +\pi$ and a boundary that does not admit the transmission of heat, so $h = 0$. In this case, it is easy to see that the eigenfunctions are $\cos mx$, $m = 0, 1, 2, \ldots$. For the solution of the cooling problem, one has the Fourier series

$$V(t, x) = \sum_{m=0}^{\infty} \phi_m(t)\cos mx.$$

The series has to be convergent and sufficiently differentiable, at least once with respect to time, twice with respect to the spatial variable. Imposing the initial condition

$$V_0(x) = \sum_{m=0}^{\infty} \phi_m(0) \cos mx,$$

we obtain the initial data $\phi_m(0)$ as Fourier coefficients, but also that $V_0(x)$ must be an even function. The Fourier series represents an even continuation of the function on the whole x-axis beyond $(-\pi, +\pi)$. At the points $\pm n\pi, n = 1, 2, \ldots$, we will in general have discontinuities. Solving the separated equation for the time-dependent part and applying the initial $\phi_m(0)$, we obtain the solution

$$V(t, x) = \sum_{m=0}^{\infty} A_m(t) \cos mx.$$

The function will in general be discontinuous near the boundary, and the expansion with respect to powers of t obtained above makes no sense. Poincaré was at this point inspired by Chebyshev [Chebyshev 1907], who had developed orthogonal polynomial expansions to solve a kinematic geometric problem but found that they did not converge. Chebyshev "solved" his problem by requiring the error of his expansion to satisfy a minimal value in the sense of least squares. In a similar way, Poincaré required the solution of the cooling problem to have a small average error $S(t)$ at time t by considering

$$S(t) = \int_D \left(V(t, x) - \sum_{m=0}^{N} A_m(t) \cos mx \right)^2 dx,$$

which becomes smaller as N increases. This convergence in the mean is now called convergence in L^2 norm.

The Fourier Integral as a General Tool

In Chapter 8 of [Poincaré 1890a, nr. 8], Poincaré discusses various features of the heat equation, the wave equation, and the telegraph equation. Consider the unbounded domain $-\infty < x < +\infty$ and $t \geq 0$. The first two equations are

$$\frac{\partial U}{\partial t} = \frac{\partial^2 U}{\partial x^2}, \qquad \frac{\partial^2 U}{\partial t^2} = \frac{\partial^2 U}{\partial x^2}.$$

The telegraph equation is

$$\frac{\partial^2 V}{\partial t^2} + 2\frac{\partial V}{\partial t} = \frac{\partial^2 V}{\partial x^2}.$$

It was developed as a model to describe the propagation of a current in a conducting medium. Introducing into the telegraph equation

$$V = Ue^{-t},$$

we have

$$\frac{\partial^2 U}{\partial t^2} = \frac{\partial^2 U}{\partial x^2} + U.$$

For the heat equation we will prescribe the initial condition $U(x,0)$; for the wave equation and the telegraph equation we also have to add the initial condition $\partial U/\partial t(x,0)$. For all three equations in U we will look for a solution of the form

$$U(x,t) = \int_{-\infty}^{\infty} \phi(q,t)e^{iqx}\, dq. \tag{11.1}$$

Starting with the heat equation, we differentiate the integral of (11.1) once with respect to t, twice with respect to x, and substitute the resulting expressions to obtain

$$\frac{\partial \phi}{\partial t} = -q^2\phi,$$

with solution

$$\phi(q,t) = \alpha(q)e^{-q^2 t}.$$

To solve the initial value problem and determine $\alpha(q)$, we have to find the function $\theta(q)$ such that

$$U(x,0) = \int_{-\infty}^{\infty} \theta(q)e^{iqx}\, dq.$$

We can find θ, and the solution of the heat equation with initial value will be

$$U(x,t) = \int_{-\infty}^{\infty} \theta(q)e^{-q^2 t}e^{iqx}\, dq.$$

In the same way, we treat the wave equation with initial values. After differentiating twice with respect to t and x, we obtain after substitution

$$\frac{\partial^2 \phi}{\partial t^2} = -q^2\phi,$$

with solution

$$\phi(q,t) = \alpha(q)\cos qt + \beta(q)\sin qt.$$

The coefficients $\alpha(q)$ and $\beta(q)$ have to be determined. Applying the initial conditions, we have to find $\theta(q)$ and $\theta_1(q)$ such that

$$U(x,0) = \int_{-\infty}^{\infty} \theta(q) e^{iqx} \, dq, \qquad \frac{\partial U}{\partial t}(x,0) = \int_{-\infty}^{\infty} \theta_1(q) e^{iqx} \, dq.$$

The solution for the initial value problem of the wave equation becomes

$$U(x,t) = \int_{-\infty}^{\infty} \left(\theta(q) \cos qt + \frac{\theta_1(q)}{q} \sin qt \right) e^{iqx} \, dq.$$

For the telegraph equation, after transforming $V \mapsto U$, we have again to determine $\theta(q)$ and $\theta_1(q)$ from the initial conditions. After differentiating twice and substituting the expression in (11.1), we obtain

$$\frac{\partial^2 \phi}{\partial t^2} + (q^2 - 1) \, \phi = 0,$$

with solution

$$\phi(q,t) = \gamma(q) \cos \sqrt{q^2 - 1} \, t + \delta(q) \sin \sqrt{q^2 - 1} \, t.$$

The coefficients $\gamma(q)$ and $\delta(q)$ have to be determined. Applying the initial conditions, we again obtain

$$U(x,0) = \int_{-\infty}^{\infty} \theta(q) e^{iqx} \, dq, \qquad \frac{\partial U}{\partial t}(x,0) = \int_{-\infty}^{\infty} \theta_1(q) e^{iqx} \, dq,$$

with $\theta(q) = \gamma(q)$ and $\theta_1(q) = \delta(q) \sqrt{q^2 - 1}$. The solution of the telegraph equation follows from

$$U(x,t) = \int_{-\infty}^{\infty} \left(\theta(q) \cos \sqrt{q^2 - 1} \, t + \frac{\theta_1(q) \sin \sqrt{q^2 - 1} \, t}{\sqrt{q^2 - 1}} \right) e^{iqx} \, dq.$$

Conclusions

For the actual solutions, we have to compute $\theta(q)$ and $\theta_1(q)$ by combining the initial conditions and the Fourier integral. Poincaré considers a number of illustrative examples. Suppose, for instance, that for the wave and telegraph equations, $U(x,0) = U_t(x,0)$, and that they are nonzero only for $b < x < a$. If the initial functions are polynomial, then $U(x,t)$ will be holomorphic in x and t except at $x = a \pm t$, $x = b \pm t$. Discontinuities will propagate along these characteristic straight lines with constant speed. For $x < b - t$ and $x > a + t$, the solution will be identically zero. This is in contrast to the heat equation, where there is a discontinuity only at $t = 0$.

An interesting example for the telegraph equation is the case that $U(x, 0)$ is identically zero, while $U_t(x, 0) = \pi/\varepsilon$ if $-\varepsilon < x < +\varepsilon$ and is identically zero outside this interval (ε is a small positive parameter). One obtains

$$\theta(q) = 0, \quad \theta_1(q) = \frac{\sin q\varepsilon}{q\varepsilon},$$

which for small ε is very close to 1. It follows that the solution is very close to

$$U(x, t) = \int_{-\infty}^{\infty} \frac{\sin \sqrt{q^2 - 1}\, t}{\sqrt{q^2 - 1}} e^{iqx}\, dq.$$

This expression can be analysed using complex contour integration. One of the conclusions is that propagation of electricity according to the telegraph equation produces a (possibly small) residual effect; this is not the case for propagation according to the wave equation in one dimension.

11.2 Rotating Fluid Masses

In the eighteenth century, scientists became interested in the equilibrium shapes of rotating fluid masses under the action of their own gravity and the fluid pressure that is present; such fluids are called self-gravitating. The interest arose from discussions about the shape of planets and stars. The underlying physical assumptions were far removed from modern insights such as internal energy production, internal motions, and dissipative effects, so the modelling is too simple for modern astrophysics. But the results and analysis are still basically of mathematical and even some astrophysical interest. A relatively recent survey of the literature can be found in [Lebovitz 1998]. For Poincaré's results we will use his lecture notes [Poincaré 1890a, nr. 12]. One of his earlier publications was in the *Acta Mathematica* for 1885. Poincaré's interest was triggered by an idea put forward by William Thomson (Lord Kelvin), who observed that our planetary system consists of many bodies and also that star systems seem to be more often multiple than not; he then made the hypothesis that such multiplicity arose from fission of rotating fluid masses. A possible mechanism for such an instability would be small friction; see also the classic monograph [Thomson and Tait 1883] or the modern survey [Kirillov and Verhulst 2010].

Regarding the stability of the new equilibrium figures found by Poincaré, his analysis is incomplete. The pear-shaped figure that he thought very promising turned out to be unstable, but some of the other solutions emerging from higher-order harmonics show interesting aspects. Paul Appell was one of the contributors to a solution of the stability problems of the pear-shaped figures; see Section 4.3 and [Appell 1921]. For developments since Poincaré's time, see [Lebovitz 1998].

In linearizing around an equilibrium state to determine its stability, one calculates the eigenvalues. If these are all purely imaginary, then the equilibrium was called "ordinary stable," while in modern times it is called "neutrally stable" or "Lyapunov stable." If all the eigenvalues have negative real parts, the equilibrium was called "secularly stable." This is now called "asymptotically stable." In the sequel we will keep to the modern terminology.

Results by Maclaurin and Jacobi

In the equations, one assumes equilibrium between gravitational forces and the fluid pressure, while overall, the Newtonian gravitational force holds the fluid together. In the fluid mass, the gravitational potential Φ is governed by the Poisson equation

$$\Delta \Phi = -4\pi \rho,$$

with $\rho(x, y, z)$ the density of the fluid. Outside the fluid, the right-hand side of the equation vanishes, producing the Laplace equation. Rotation with constant angular velocity ω is expected to produce axisymmetric figures that are flattened at the poles. Supposing that the rotation takes place around the z-axis, one introduces the reduced potential U by

$$U = \Phi + \frac{\omega^2}{2} \left(x^2 + y^2 \right).$$

If $p(x, y, z)$ is the pressure, we have at equilibrium a balance of forces given by

$$\frac{\partial p}{\partial x} = \rho \frac{\partial U}{\partial x}, \quad \frac{\partial p}{\partial y} = \rho \frac{\partial U}{\partial y}, \quad \frac{\partial p}{\partial z} = \rho \frac{\partial U}{\partial z}.$$

There are various expressions for the energy of the system; putting

$$W = \frac{1}{8\pi} \int \left(\left(\frac{\partial \Phi}{\partial x} \right)^2 + \left(\frac{\partial \Phi}{\partial y} \right)^2 + \left(\frac{\partial \Phi}{\partial z} \right)^2 \right) dx \, dy \, dz,$$

and for the inertial moment with respect to the z-axis

$$J = \int \rho(x, y, z) \left(x^2 + y^2 \right) dx \, dy \, dz,$$

we have that stationary points of the functional

$$W + \frac{\omega^2}{2} J$$

correspond to relative equilibria. If the stationary point corresponds to a maximum, the equilibrium is Lyapunov stable. Without rotation, the sphere gives the stable equilibrium solution.

Maclaurin (1698–1746) derived an explicit expression for a rotating fluid mass that is an oblate spheroid. The flattening (eccentricity) depends on ω. The fluid mass is assumed to be in solid rotation. Later, Maclaurin's results were generalized to ellipsoids with all axes unequal, the so-called triaxial Maclaurin ellipsoids.

Jacobi (1804–1851) assumed a slightly simpler expression for the potential and found a second family of triaxial ellipsoids that is independent of Maclaurin's ellipsoids. Interestingly, there is one point in parameter space where they coincide, a bifurcation point of the families of ellipsoids.

Both families represent special solutions of solid mass rotation with corresponding density distributions, eccentricities, and rotational velocities.

Ellipsoids with Internal Dynamics

Important steps forward were made successively by Dirichlet (1805–1859), Dedekind (1831–1916), and Riemann (1826–1866). They began with the partial differential equations of fluid mechanics to find solutions by a similarity approach. Assuming a certain spatial structure produces equations for functions of time. Dirichlet was able to solve the equations in the case of a homogeneous ellipsoid. Dedekind gave more details of Dirichlet's model and added an ellipsoid that is characterized by motions of constant vorticity. Riemann, in turn, clarified these models and added stability considerations.

Poincaré's Contribution

Thomson's fission hypothesis was to consider the evolution of steady-state solutions, such as the rotating Maclaurin ellipsoids with increasing rotational velocity, that at some stage of evolution would split into two equilibrium figures. The mechanism to produce such a bifurcation could be rotational or dissipation-induced instability caused by the small viscosity of the fluids. Among the scientists who studied this scenario assuming solid-body rotation were Lyapunov and Poincaré; the results on ellipsoids with internal dynamics were largely ignored.

In [Poincaré 1890a, nr. 12], the theory is developed from first principles, starting with Newtonian gravitational attraction, defining the gravitational potential Φ, and stating the usual basics of potential theory—the Laplace equation, the Poisson equation, and the theorems of Gauss and Green. A homogeneous fluid is considered, following the analysis by Maclaurin and Jacobi. In this case, the constant angular velocity ω has an upper limit, and in suitable physical units, a necessary condition for equilibrium is

$$\omega^2 \leq 2\pi.$$

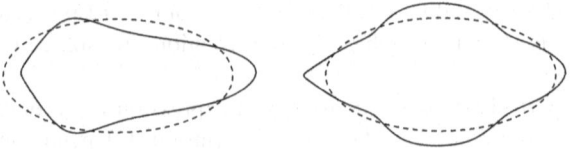

Fig. 11.3 The pear-shaped (left) and other figures branching off Jacobi's ellipsoids. They are associated with Lamé functions, higher-order harmonics in ellipsoidal coordinates. A dotted ellipse indicates a cross section of an unperturbed ellipsoid

For the two types of triaxial ellipsoids attributed to Maclaurin and Jacobi, Poincaré derives the conditions $\omega < 4\pi \times 0.112$ and $\omega < 4\pi \times 0.093$.

According to Dirichlet, for an equilibrium to be stable, the expression

$$W + \frac{\omega^2}{2} J$$

has to be maximal. William Thomson observed that the equilibrium is in this case also asymptotically stable if we add friction. If an equilibrium does not maximize the expression, it can be Lyapunov stable, but in that case it may be unstable with respect to dissipative effects. Poincaré notes that in the case of solid rotation, we have no friction. He shows also that in the case of triaxial ellipsoids, stable equilibrium requires rotation about the smallest axis.

In Poincaré's lecture notes, considerable attention is paid to suitable orthogonal special functions. These are the spherical functions, polynomials derived in polar coordinates from the Laplace equation; and the Lamé functions, polynomials derived in ellipsoidal coordinates. The Lamé functions play a prominent role in the expansions for Maclaurin's and Jacobi's triaxial ellipsoids. At this point, using higher-order Lamé functions, Poincaré discovered a new series of solutions branching off the Jacobi ellipsoids. He called them "pear-shaped"; see Figure 11.3.

If the constriction of the pear-shaped figures narrowed during evolution, Poincaré expected cooling of the fluid; such slowly evolving equilibria would be suitable candidates for the fission theory of planetary and stellar systems. His stability analysis depended on linearization of the equations and was not completed during his lifetime.

An interesting result that plays a part in general bifurcation theory is the phenomenon of "exchange of stabilities." For instance, in the so-called transcritical bifurcation, two solutions exist depending on whether a certain parameter μ is less or greater than a critical value. At the critical value, they coincide; one of the solutions is stable, the other unstable, and this characteristic is exchanged when the critical value is passed. A simple example is the equation

$$\dot{x} = \mu x - x^2,$$

with critical value $\mu = 0$; see Figure 9.10.

Another interesting feature of the lecture notes is Poincaré's analysis of the rings of Saturn. There are three possibilities for the rings: they are solid, liquid, or particulate. A solid ring turns out to be unstable for physically realistic values of the parameters. For a liquid ring, Poincaré computes the gravitational potential of a rotating torus. For stability, a necessary condition is a very low density of the ring, while on the other hand, the pressure should be sufficiently large. These requirements are incompatible. What remains is the possibility of a ring of particles that are separated from each other; this can be a stable configuration.

11.3 Dynamics of the Electron: Poincaré Group and Relativity

The Larmor Papers of 1895

In 1895, Poincaré wrote a series of four articles on a theory propounded by Joseph Larmor (1857–1942); see [Poincaré 1895a] and [Poincaré 1916, Vol. 9]. It contains a lucid discussion of the theory and experiments regarding optics and electricity, referring to Larmor, Fresnel, Lorentz, Helmholtz, and Hertz. It is of interest that his ideas at that time already showed a preparation for the theory of special relativity.

In the preliminary conclusions [Poincaré 1916, Vol. 9, pp. 409–413], he states that none of the present theories combines the theoretical requirements of consistency and an explanation of the experiments; the best one seems to be the theory of Lorentz, see Figure 11.4. Altogether, this is an unsatisfactory state of affairs. He concludes in 1895 with a revolutionary statement:

> The experiments have produced a host of facts that can be summarized in the following form: it is impossible to demonstrate the absolute motion of matter, or better formulated, the relative motion of substantial matter with respect to the ether; what can be made evident is the motion of substantial matter with respect to substantial matter.

Referring to an experiment by Michelson, he adds that not only is it impossible to demonstrate any motion of matter with respect to the ether, but moreover, the problem of incorporating the law of "action is reaction" (Newton's third law) in a description of interaction between matter and ether is unsolved. He concludes that both facts have to be related to each other.

The Zeeman Effect

Poincaré was very much interested in new experiments in physics. One such experiment demonstrated the so-called Zeeman effect, which is the splitting of spectral lines in radiation emitted by an atom that is placed in a magnetic field. He continued his reflections on physical phenomena in 1897 and 1899 with a description of the Zeeman effect in the context of Lorentz's theory. (see [Poincaré

Fig. 11.4 Hendrik Antoon
Lorentz, physicist

1916, pp. 427–460]). The analysis is of historical interest, but it was superseded
by the developments of quantum mechanics that replaced parts of classical physics
around 1900.

The Paper in Honour of Lorentz: 1900

On December 11, 1900, a celebration took place at the university of Leiden on the
occasion of the 25th anniversary of Hendrik Antoon Lorentz's being awarded his
doctoral degree. Poincaré used this occasion to discuss again the problem of the
law of "action is reaction" in electrodynamics (published as [Poincaré 1895b]). He
modified earlier remarks that had called Lorentz's theory the best available but were
still mildly critical:

> Without doubt one will find it strange that in a monument erected for the glory of Lorentz,
> I return to considerations presented earlier as an objection to his theory. I could say that
> the pages that follow are more softening than aggravating the objection. But I find this
> not a good excuse because I have one that is a hundred times better: *good theories are
> flexible*. Those that have a rigid form that cannot be removed without falling to pieces, have
> really very little vitality. But if a theory shows us some true relations, it can be clothed in a
> thousand different forms, it will resist all attacks and its essence will not change.

In considering the forces on a collection of electrons bounded in a certain volume
and to satisfy the "action is reaction" law, Poincaré has to assume the presence of
nonelectric forces. One of the consequences is this [Poincaré 1916, p. 471]:

> Since the electromagnetic energy behaves in our point of view like a fluid endowed with
> inertia, we have to conclude that a device, after having produced electromagnetic energy,

radiates in a certain direction; the device has to recoil as a cannon has to recoil when it has launched a projectile.

At this point, Poincaré could have made the step of equating electromagnetic energy with mass, but that step was too big. A numerical example of the phenomenon shows that the recoil effect is small and difficult to observe. For further understanding, one has to consider the motions as relative. If in one dimension, a particle has position x with respect to the observer and v is its velocity in a moving frame of reference, then the position in a reference frame indicated by x' satisfies the relation

$$x' = x - vt.$$

But according to Lorentz, we have to introduce *local time t'* by the transformation

$$t' = t - \frac{vx}{c^2}.$$

The explanation for the need of local time is given and repeated extensively in Poincaré's Göttingen lecture; see Section 11.4.

The constant c is the velocity of light. In the reference frame, the local time is a second-order effect with respect to $1/c$. For the relative motion in the reference frame, the total energy is not equal to the energy observed at position x. It appears that an additional force acts in the reference frame. This looks like a contradiction, but we have to conclude that the energy radiated by a device at the position of the observer is not equal to the energy—in fact, it is greater—radiated by a device placed in a moving frame. The apparent radiation and the apparent recoil energy will make up the difference. It is thus that the principle of "action is reaction" in Lorentz's theory can be interpreted.

The principle of "action is reaction" is fundamental in physics, so it is not surprising that the discussion of the principle went on with contributions by Abraham, Planck, Lorentz, and others. For references, see [Poincaré 1916, Vol. 9, p. 698.].

The Dynamics of the Electron: 1905–1906

A *Comptes Rendus* paper [Poincaré 1905a] of 1905 and its longer version of 1906 [Poincaré 1906] are concerned with the dynamics of the electron. Another description and comments can be found in [Le Bellac 2010].

The paper [Poincaré 1905a] was submitted on June 5, 1905, which means that it was submitted earlier than Einstein's famous paper on special relativity. The five pages announcing [Poincaré 1906] contain the following ideas:

1. All physical experiments show that the impossibility of demonstrating the absolute motion of matter is a general law of nature.

2. Lorentz has proposed to explain this in [Lorentz 1904] by the contraction of moving bodies. It explains the present experiments and asks for testing against new experiments.

3. The Lorentz transformation contains several parameters, among which is a multiplicative factor function $l(\varepsilon)$. The motion is in the x-direction; ε indicates the ratio of the velocity of the body to the velocity of light, $\varepsilon = v/c$, and we have with $k = 1/\sqrt{1 - \varepsilon^2}$,

$$x' = kl(x + \varepsilon t), \quad y' = ly, \quad z' = lz, \quad t' = kl(t + \varepsilon x);$$

they form a group of transformations. Lorentz gives some arguments for setting $l = 1$, but Poincaré gives a conclusive argument in [Poincaré 1905a] that the transformation should have rotational invariance, which implies $l = 1$.

4. When in motion, an electron can be deformed and compressed as if an exterior force were acting.

5. In applying Lorentz's transformation to all forces of nature, one should conclude that the propagation of a gravitational force has to be with the velocity of light. The gravitational attraction of a moving body should take into account the position and velocity of the body to determine the emitted gravitational wave. The difference with Newton's gravitational law is expected to be inversely proportional to the square of the velocity.

There are various statements by Poincaré on the *principle of relativity*. One of them is in [Poincaré 1906]; also in [Poincaré 1916, p. 495]:

Il semble que cette impossibilité de mettre en évidence expérimentalement le mouvement absolu de la Terre soit une loi générale de la Nature; nous sommes naturellement porté à admettre cette loi, que nous appellerons le *Postulat de Rélativité* et à l'admettre sans restriction.

(It seems that this impossibility of establishing experimentally the absolute motion of the Earth is a general law of nature; we are, of course, set towards admitting this law that we will call the *Postulate of Relativity* and to admit it without restriction.)

The first part of [Poincaré 1906] (submitted July 23, 1905) is concerned with the analysis of the Lorentz transformation for a given coordinate system. Consider an electron as a moving body in the x-direction and a small sphere around the electron. The comoving sphere is described by the equation

$$(x - \xi t)^2 + (y - \eta t)^2 + (z - \zeta t)^2 = r^2.$$

The Lorentz transformation changes this sphere into an ellipsoid, and if the electron's charge is invariant, the electrical charge density ρ' becomes

$$\rho' = \frac{k}{l^3}(\rho + \varepsilon \rho \xi).$$

The continuity equation

$$\frac{\partial \rho'}{\partial t'} + \sum \frac{\partial \rho' \xi'}{\partial x'} = 0$$

has been satisfied. This result differs slightly from Lorentz's density ρ' in [Lorentz 1904]. In the same way, Poincaré derives the new electric and magnetic field expressions, and in addition, the forces acting in the moving frame. Here again, there is a difference with Lorentz's expressions. The equations for the electric and magnetic fields are the same, but the equation for the moving ellipsoid is not. How is one to explain this difference?

To derive the transformations, one has to use the principle of relativity and the minimization of a functional ("le principe de moindre action"). This variational approach leads to Lorentz's expressions for apparent positions and time, but eventually to Poincaré's results described above.

The announcement in [Poincaré 1905a] that the Lorentz transformations form a group is worked out in [Poincaré 1906, Section 4]. The group of transformations, denoted by ϕ, admits the following:

1. The dilation T_0, permutable to the other transformations:

$$T_0 = x \frac{\partial \phi}{\partial x} + y \frac{\partial \phi}{\partial y} + z \frac{\partial \phi}{\partial z} + t \frac{\partial \phi}{\partial t}.$$

2. The boosts T_1, T_2, T_3 acting along the respective axes. For instance,

$$T_1 = t \frac{\partial \phi}{\partial x} + x \frac{\partial \phi}{\partial t}.$$

3. The rotations $[T_1, T_2]$, $[T_2, T_3]$, and $[T_3, T_1]$.

Combinations of these transformations are permitted and result in a linear transformation conserving the quadratic form

$$x^2 + y^2 + z^2 - t^2. \tag{11.2}$$

By putting $y_4 = it$ as the fourth coordinate, Poincaré introduces [Poincaré 1916, Vol. 9, p. 542] the metric

$$ds^2 = dy_1^2 + dy_2^2 + dy_3^2 + dy_4^2.$$

This is the metric also introduced by Minkowski in 1908.

In the Lie group we can apply the commutator of the infinitesimal generators. If we turn the system through an angle π around the y-axis, the transformation becomes

$$x' = kl(x - \varepsilon t), \quad y' = ly, \quad z' = lz, \quad t' = kl(t - \varepsilon x).$$

However, in a group, the inverse transformation

$$x' = \frac{k}{l}(x - \varepsilon t), \quad y' = \frac{y}{l}, \quad z' = \frac{z}{l}, \quad t' = \frac{k}{l}(t - \varepsilon x),$$

should be identical. It follows that $l = 1/l$, or $l = 1$ ($l = -1$ produces a physically equivalent solution).

The formulation of the Lorentz group and the establishing of $l = 1$ belong to the permanent results of the paper on the dynamics of the electron. In the sequel, Poincaré links the results with Langevin waves in an electromagnetic field and the apparent deformation of electrons in a moving frame. As expected, it is impossible [Poincaré 1906, Sections 7 and 8] to use observations of this apparent deformation to establish an absolute frame of reference for motion.

The consequences for the theory of gravitation are analysed in the last section of [Poincaré 1906]. The change with respect to the classical theory will be that gravitation will depend not only on position and mass, but also on the velocity of the mass at time t_0. In addition, since gravitation takes time to travel, we have to take into account the position and velocity at time $t_0 + t$ when the propagation of the force began; so t will be negative. There are natural conditions such as satisfying the Lorentz transformations, the reduction to the classical gravitational laws if the bodies are at rest, finding small deviations from the classical laws if the velocity of the moving body is small with respect to the velocity of light. However, these conditions are insufficient to determine the laws of gravitation in a new framework. What one can consider is the case of two bodies moving with the same velocity. One can write down the Lorentz transformations for this case, but this does not determine t. Progress can be made by introducing the invariants of the Lorentz group; the substitutions of the group do not change the quadratic form (11.2). Points in space have coordinates $x, y, z, \sqrt{-1}\,t$, and in applying the coordinate transformations and the invariants, the only consistent choice that does not lead to contradictions is that gravitation propagates with the velocity of light. Since the deviations from Newton's laws are quadratic in the ratio of velocity of the body and the velocity of light, the effect will be difficult to observe.

Final Reflections: 1912

In the last years of his life, Poincaré continued to participate in discussions about relativity, the kinetic theory of gases, and the emerging quantum theory. He gave a lecture on relativity in Göttingen (Section 11.4) for a general scientific audience. He attended the famous Solvay conference (1911) on theoretical physics in Brussels, where he met Einstein and discussed quantum theory. On April 11, 1912, a few months before his death, he gave the closing lecture at a physics conference. He chose the title "The relations between matter and ether," using the occasion to summarize a number of important physics questions of that time; see [Poincaré 1912a].

Starting with the observation that atomic theory was originally invented by Democritus, he notes that both qualitatively and to a great extent quantitatively, the theory has become quite solid. The old idea was that the atom was the indivisible smallest unit of matter, but in modern times, a world of electrons, a positive nucleus, and associated radiation has been discovered. He adds about the results of modern science:

> Democritus would have opined that after having gone to so much trouble to find them, we have made no more progress than when we started. These philosophers are never satisfied.

Poincaré notes that each discovery invokes new ones. In particular, the line spectra that have been found have to be understood. A possibility is that an atom in vibration with electrons associated with a magnetic field can be in different states according to the number of "magnetons" present. This would produce a discrete spectrum, but we have to explain what a magneton is and how an atom can have a different number of them. Part of the explanation can come from Planck's theory of quanta, which rule the exchanges between matter and ether or, differently formulated, the exchanges between matter and the small resonators that produce light by sudden jumps. At this stage, the various theories to explain the new experiments are contradictory, for instance explaining the changes of the wavelength of light by moving mirrors using quantum jumps or continuously. The phenomena observed should agree with the principle of relativity, with probability, and with the laws of thermodynamics, but it is not yet clear how this can be achieved.

11.4   The Six Lectures at Göttingen: 1909 (Relativity)

The six lectures [Poincaré 1909], altogether 60 pages, have the following titles:

1. Über die Fredholmschen Gleichungen (on the Fredholm equations).
2. Anwendung der Theorie der Integralgleichungen auf die Flutbewegung des Meeres (application of the theory of integral equations to tidal motion of the ocean).
3. Anwendung der Integralgleichungen auf Hertzsche Wellen (application of integral equations to Hertz waves).
4. Über die Reduktion der Abelschen Integrale und die Theorie der Fuchsschen Funktionen (on the reduction of abelian integrals and the theory of Fuchsian functions).
5. Über transfinite Zahlen (on transfinite numbers).
6. La mécanique nouvelle (the new mechanics).

The first three lectures deal with Fredholm equations and their applications; they are typical for the beginning of spectral theory.

In the last lecture, lecture 6, Poincaré discusses relativity while still giving a place in the theory to the ether. It is a slightly updated and shortened version of the text in [Poincaré 1908a]. One should note that he uses the term "electron" both for the negatively charged particle, as we do today, and for the positively charged nucleus. In [Poincaré 1908a], he calls the nucleus the "central electron" and the "positive electron." At the end of the lecture, he takes a rather pessimistic view of the future teaching of mechanics, expecting that teachers will feel conflicted between the old and new mechanics, but we can say that after roughly a century of mechanics teaching since 1909, this has not happened. Here follows a translation of lecture 6.

The New Mechanics

Ladies and gentlemen: Today I am obliged to speak French, and I have to excuse myself for this. It is true that in my preceding lectures, I have expressed myself in German, in very bad German: you see, speaking in foreign languages is like walking when one is lame; it is necessary to have crutches. My crutches were until now the mathematical formulas, and you have no idea what a support they are for an uncertain lecturer. In tonight's lecture, I don't want to use formulas; I am without crutches, and this is why I have to speak French.

In this world, you know, nothing is definitive, nothing is unmovable. The most powerful and most solid empires do not last forever: this is a theme that preachers often liked to develop. Scientific theories are like empires; they are not certain of the future. If one of them appeared to be safe from the wear and tear of time, it was certainly Newtonian mechanics: it seemed beyond dispute; it was an impregnable monument, and look, in its turn, I will not say that the monument has been pulled down, that would be rash, but in any case, it is strongly shaken. It has suffered attacks of great demolishers; there is one present here, Max Abraham; another is the Dutch physicist Lorentz. I would like in a few words to talk to you about the ruins of this old structure and of the new building that one wishes to erect in its place.

To start with, what characterized the old mechanics? It was this very simple fact: Consider a body at rest that undergoes a forcing, which means that during a given interval of time, I let a given force act on it. The body starts moving and obtains a certain velocity. While the body is moving with this velocity, let the same force again act on it during the same interval of time, and the velocity is doubled; if we still continue, the velocity will be tripled after we have applied the same forcing a third time. On starting in this manner a sufficient number of times, the body will obtain a very large velocity that could pass any limit, an infinite velocity.

In the new mechanics, on the other hand, one supposes that it is impossible to give a velocity to a body starting at rest that is larger than the velocity of light. What happens? I consider the same body at rest. I give it a first forcing, the same as before, and it will assume the same velocity. Resuming this forcing a second time, the velocity will still increase, but it will not be doubled. A third forcing will produce an analogous effect: the velocity increases, but less and less; the body offers a resistance that becomes bigger and bigger. This resistance is the inertia, what one usually calls the mass. So, everything that happens in this new mechanics is as if the mass were not constant but grew with the velocity. We can represent the phenomenon graphically: in the old mechanics, the body assumes after the first forcing a velocity represented by the segment Ov_1; after the second forcing, Ov_1 increases with an equal segment v_1v_2; at each new forcing, the velocity increases by the same amount; the segment representing it increases with constant length. In the new mechanics, the velocity segment increases with segments $v'_1v'_2$, $v'_2v'_3, \ldots$ that are smaller and smaller and in such a way that we cannot pass a certain limit, the velocity of light.

How has one been induced to draw such conclusions? Have direct experiments been carried out? The deviations are produced only for bodies moving with large velocities; only then are the indicated differences observable. But what is a very large velocity? Is it a car with speed 100 km/hour? In the streets one is excited about such a velocity. From our point of view it is still small, a snail's speed. Astronomy does better: Mercury, the fastest of the celestial bodies, runs at 100 km, but not in an hour, rather in a second. Still, this is not sufficient; such velocities are too weak to reveal the differences that we would like to observe. I will not talk about cannonballs; they are faster than cars but much slower than Mercury. You know, however, that one has discovered an artillery whose projectiles are much faster. I want to discuss radium, which sends energy, projectiles, in all directions. The speed of the shots is much larger, the initial velocity is 100 000 km per second, one-third the velocity of light. It is true that the calibre of the projectiles, their weight, is weak, and we must not count on this artillery to increase the military force of our armies. Can one experiment with these projectiles? One has indeed attempted such experiments. Under the influence of an electric or magnetic field, a deviation arises that enables us to take into account and to measure the inertia. In this way, one has established that the mass depends on the velocity, and one expresses it in this law: the inertia of a body grows with its velocity, which remains below the velocity of light, 300 000 km per second.

I will proceed now to a second principle, the principle of relativity. Suppose an observer moves to the right; for him everything happens as if he were at rest; the objects around him are moving to the left. He has no means to find out whether the objects are changing places, whether the observer is motionless or moving. One teaches in all courses of mechanics that the passenger on a boat thinks that he sees the borders of the river moving while he is slowly carried along by the motion of the vessel. If you examine this simple notion closely, it acquires an enormous importance. One has no means to solve the question; no experiment can falsify the principle: there is no absolute space; all the displacements that we can observe are relative displacements. Since these considerations are quite familiar to philosophers, I have occasionally expressed them. In this way I have even received publicity that I could gladly do without; all the reactionary French newspapers had me demonstrating that the Sun moves around the Earth. In the famous trial between the Inquisition and Galileo, Galileo was completely wrong.

Let us return to the old mechanics that admitted the principle of relativity. Instead of being founded on experiment, its laws were derived from this fundamental principle. These considerations were satisfactory for purely mechanical phenomena, but the old mechanics didn't work for important parts of physics, for instance optics. One considered the velocity of light relative to the ether as absolute. This velocity could be measured; theoretically, one had the means to compare the displacement of a moving object with an absolute displacement, the means to decide whether a body was in absolute motion.

Delicate experiments, with extremely precise instruments that I will not describe to you, have made it possible to do such a comparison in practice: the results have been zero. There is no restriction on the principle of relativity in the new mechanics; it has, if I dare to say it, an absolute value.

To understand the part that the principle of relativity plays in the New Mechanics, we have first to talk about a very ingenious invention of the physicist Lorentz, "apparent time." Suppose we have two observers. One is at A in Paris; the other B is in Berlin. A and B have identical stopwatches that they want to synchronize. But they are more precise observers than can be found anywhere. They require their synchronization to be of an extraordinary precision, for instance not by a second but by a billionth of a second. How are they going to do this? A sends a telegraphic signal from Paris to Berlin, wireless if you wish to be altogether modern. B perceives the moment of reception, and this will be for the two stopwatches the start of the chronology. But the signal needs a certain time to go from Paris to Berlin; it will go with the velocity of light, so B's watch will lag behind. B is too intelligent to ignore this; he will remedy this inconvenience. The matter seems quite simple; one exchanges signals; A receives and B sends, and taking the average of the corrections made in this way, one has the exact time. But are we certain of that? We are supposing that the signal uses the same time in going from A to B as from B to A. But A and B are carried along by the movement of the Earth with respect to the ether, the medium of the electrical waves. Once A has sent his signal, he runs away from it; B moves the same way, and the time needed will be much longer than if the observers were at rest. If, on the other hand, B sends and A receives, the time is much shorter, since A moves with the signals. It is absolutely impossible for them to know whether their stopwatches indicate the same time. Whatever method is used, the inconveniences remain the same; observation of an astronomical phenomenon or an optical method suffers from the same difficulties. B will never know more than an apparent difference of time, a kind of local time. The principle of relativity applies in all respects.

Nevertheless, in classical mechanics one demonstrated with this principle all fundamental laws. Could one be tempted to reconsider the classical reasoning and reason like this? Let's have again two observers, call them A and B, as they are usually called in mathematics. Assume that they are moving away from each other; neither of them can exceed the velocity of light. For instance, B moves at 200 000 km/sec to the right, A at 200 000 km/sec to the left. A can imagine being at rest, and the apparent velocity of B will be for him 400 000 km/sec. If A knew the new mechanics, he would say: B has a velocity that he cannot reach, so I am also moving. It seems that he could reach an absolute decision about his situation. But he himself must be able to observe the position of B. To make this observation, A and B start by synchronizing their watches, after which B cables A to indicate his successive positions. Collecting them, A makes an account of the motion of B, and he can outline the curve of the motion. But now the signals are propagating at the velocity of light; the watches that indicate the apparent time are varying at each moment, and everything happens as if B's watch were fast. B will believe that he is going much slower, and the apparent velocity with respect to A will not pass the limit that it should not reach. Nothing can reveal to A whether he is in motion or at absolute rest.

One has to make a third hypothesis that is much more surprising, much more difficult to accept, since it conflicts with our actual experiences. A body moving in translation undergoes a deformation in the direction of displacement. A sphere, for instance, becomes a kind of flattened ellipsoid whose minor axis will be parallel to the translation. If one does not observe such a transformation every day, it is because its smallness makes it nearly undetectable. The Earth carried along in its orbit deforms by about a factor $1/200\,000\,000$. To observe such a phenomenon, one needs measuring instruments with extreme precision, but their precision would have to be infinite, and we would not get any further, since they are also carried along in the motion undergoing the same transformation. We will observe nothing; the measuring rod that we could use will become as short as the length to be measured. One can only require some knowledge when comparing the length of one of these bodies at the velocity of light. These are delicate experiments, carried out by Michelson. I will not show you the details; they have produced very remarkable results. Strange as they seem to us, we have to admit that the third hypothesis has been verified perfectly.

The foundations of the new mechanics are such that with application of these hypotheses, they are compatible with the principle of relativity.

But it must still be connected with a new concept of matter. For modern physics, the atom is not a simple element; it has become a real universe in which thousands of planets gravitate around very small suns. Suns and planets are here electrically charged particles, negative or positive. The physicist calls them electrons, and he establishes the world with them. Nobody would have imagined that the neutral atom is a central positive mass around which a large number of negatively charged electrons are revolving with total electrical charge equal in size to that of the central nucleus. This concept of matter makes it easier to account for one of the characteristics we have outlined for the new mechanics, the increase of the mass of a body with its velocity. Since an arbitrary piece of matter is nothing but a collection of electrons, it suffices to demonstrate this for them. For this purpose, we note that an isolated electron that moves through the ether brings along an electrical current, implying an electromagnetic field. This field corresponds to a certain quantity of energy that is not localized in the electron, but in the ether. A variation in size or in direction of the velocity of the electron modifies the field and expresses itself by a variation of the electromagnetic field of the ether. Whereas in Newtonian mechanics the use of energy is caused by the inertia of the moving body, here a part of this use is caused by what one could call the inertia of the ether with respect to the electromagnetic forces. The inertia of the ether increases with the velocity and becomes infinite in its limit when the velocity tends near the velocity of light. So, the apparent mass of the electron increases with the velocity; the experiments of Kaufmann show that the real constant mass of the electron is negligible with respect to the apparent mass and can considered to be zero.

In this new concept, the constant mass of matter has vanished. Only the ether, and no longer matter, is inert. Only the ether puts up resistance to motion so that we could say, there is no matter, there are only holes in the ether. For stationary or quasistationary motion, the new mechanics does not diverge—taking into account the level of approximation of our measurements—from Newtonian mechanics, with the only difference that the mass is not independent of the velocity, nor the angle that this velocity makes with the direction of the force of acceleration. If, on the other hand, the velocity has a considerable acceleration, in the case, for instance, of very fast oscillations, there is a production of Hertz waves that represent a loss of energy of the electron, which undergoes damping of its motion. In this way, in wireless telegraphy, the emitted waves are caused by oscillations of the electrons in oscillatory discharge.

Analogous vibrations take place in a flame and even in a white-hot body. According to Lorentz, inside a white-hot body, a considerable number of electrons are circulating that fly in all directions; since they cannot leave the body, they reflect against its surface. One could compare them to a cloud of small insects enclosed in a bowl that are beating with their wings against the boundaries of their prison. The higher the temperature, the faster the motion of these electrons, and there will be more mutual collisions and reflections against the boundary. At each collision and each reflection, an electromagnetic wave is emitted, and it is the observation of these waves that makes the body appear to be white-hot.

The motion of the electrons is nearly detectable in a Crookes tube. It produces a real bombardment of electrons leaving the cathode. These cathode rays hit the anticathode violently, and as they reflect, a portion of them gives birth to an electromagnetic shock that several physicists identify with Roentgen rays.

To conclude, it remains to examine the relations between the new mechanics and astronomy. As the notion of constant mass of a body disappears, what will become of Newton's law? It can persist only for bodies at rest. Moreover, one needs to take into account that attraction is not instantaneous. It is reasonable to ask oneself whether the new mechanics will not result in making astronomy more complicated without obtaining a superior approximation to that given by classical celestial mechanics. Mr Lorentz has considered the question. Starting with the assumption that Newton's law is correct for two electrically charged bodies at rest, he calculates the electrodynamic effect of the currents produced by these bodies in motion. In this way, he obtains a new attraction law containing the velocities of the two bodies as parameters. Before examining how this law accounts for astronomical phenomena, we want to note in addition that the acceleration of celestial bodies has as an outcome electromagnetic radiation, so dissipation of energy will make itself felt by damping of their velocity. So, in the long run, the planets will fall into the Sun. But this perspective cannot scare us, since the catastrophe cannot arise within several multibillions of centuries. Returning now to the law of attraction, we can easily see that the difference between the two mechanics will be larger as the velocities of the planets are larger. If there is a notable difference, it will be the greatest for Mercury, since Mercury has the largest velocity of all the planets. But as it happens, Mercury presents precisely an anomaly that has not yet been explained:

its perihelion motion is much faster than the motion calculated by the classical theory. The acceleration is 38 seconds too large. Le Verrier attributed this anomaly to a not yet discovered planet, and an amateur astronomer thought he had observed its transit of the Sun. Since that time, nobody has seen it, and it is unfortunately certain that this observed planet was only a bird. Now the new mechanics accounts for the direction of the relative error of Mercury, but it leaves nonetheless a deviation of 32 seconds with observation. So it does not suffice to bring agreement between the theory and observation of Mercury. Although this result is not decisive in favour of the new mechanics, it is not opposed to its acceptance, since the direction in which it corrects the deviation of the classical theory is the right one. The theory of the other planets is not noticeably modified in the new theory, and the results agree approximately with measurements and classical theory.

To conclude, I believe it would be premature to consider classical mechanics as definitively finished, notwithstanding the great value of arguments and facts raised against it. Whatever happens to it, it will remain the mechanics of very small velocities with respect to the velocity of light, the mechanics of our life in practice and our earthly technology. If, however, its rival triumphs within a few years, I will permit myself to draw your attention to a pedagogical stumbling block that many teachers will not escape, at least in France. When teaching elementary mechanics to their students, these teachers would feel strongly that teaching them this mechanics was past its time, that a new mechanics in which the notions of mass and time have a very different meaning will replace it. They will regard from on high this out-of-date mechanics that the curriculum forces them to teach, and they will let their students feel their disdain for it. Nevertheless, I believe that this despised classical mechanics will be as necessary as it is now and that those who do not know it well will not understand the new mechanics.

11.5 Cosmogony

In his lecture notes on cosmogony [Poincaré 1890a, nr. 14], Poincaré takes the modern view that the solar system can be understood only from the point of view of evolution. The part of the universe containing the solar system has clearly developed from chaos to order. This was caused by energy exchanges, dissipation processes such as tidal friction, and other mechanisms.

The book appeared around a hundred years ago, and in it, Poincaré does not advance much new cosmogonic theory, and since that time, our empirical and theoretical knowledge of the solar system has increased tremendously. Is the book still of interest? The clear, analytical discussion of the cosmogonic hypotheses of Laplace and Darwin makes it so. The development of the virial theorem, the use of adiabatic invariants, and elements of statistical mechanics are original contributions. The astrophysical discussions of the later chapters are mainly of historical interest, and we will only briefly mention some of those aspects.

Theories regarding the origin of the solar system should explain a number of phenomena. The main points around 1910 were these:

1. The planets are moving around the Sun in nearly circular orbits and nearly in a plane (coplanarity).
2. Most of the mass is contained in the Sun, most of the angular momentum in the motion of the planets.
3. A mechanism for star and planet formation should explain the mass distribution and distances in the solar system (four relatively small inner planets, massive outer planets).
4. The presence of asteroids, comets, and debris in the solar system.
5. The presence of some of the bodies in retrograde motion ("direct" rotation follows the rotation of the Sun and the motion of the Earth around the Sun; "retrograde" means rotation in the opposite direction).

At the present time, highly improved measurement techniques and the development of astrophysics have added to these requirements. We mention the following:

1. General insight into star formation and the evolution of the Sun.
2. More precise masses and densities of the planets.
3. Information on the composition of the planets, satellites, and small bodies in the solar system.
4. Detailed information about the dynamics of the planets, satellites, and small bodies.
5. Observations of planetary systems outside the solar system.

We review the contents of the book with, as announced, emphasis on the nebula hypothesis of Laplace and the tidal friction theory of Darwin.

Hypotheses around the Original Nebula

The philosopher Immanuel Kant (1724–1804) wrote about the origin of the solar system in 1755 and 1763. As described in Chapter 1, Kant assumes in the beginning a homogeneous medium at rest, in which condensations form, a large one in the centre, leading to planets revolving around the Sun. Starting with a medium at rest, this circulation is not in agreement with the conservation of angular momentum (the law of areas). According to Kant, the density of Sun and planets have to be the same, and all rotation has to be direct. Observations of comets in retrograde motion have to be erroneous. The picture sketched by Kant is qualitative only and in contradiction with the laws of mechanics.

Laplace (Chapter 2) does not start at the formation of the solar system, but assumes the presence of a central condensation with a very large nebulous atmosphere. The condensation is, together with the nebula, in rotation; the boundary of the nebula is determined by the balance of centrifugal forces and pressure. Near the central condensation, molecules were attracted, leading to the formation of the

Sun; outside, rings of matter were formed in which by shocks and gravitational contraction the planets were shaped. The cooling of the nebula together with the instability of the rings produced the planetary system. The only rings left of the original state are the rings of Saturn. The comets are often in retrograde motion and have sizable eccentricities and inclinations. Laplace considers the comets as bodies that originated from other planetary systems. His theory is both qualitative and quantitative.

Chapter 3 contains an extensive analysis of the hypotheses of Laplace, partially based on the work of Edouard Roche (1820–1883). Assuming a uniform angular rotation ω and central mass M, one can write down the equation for the equipotential surfaces and derive the shape and size of the nebula. If the nebula contracts because of cooling, ω will increase (conservation of angular momentum). The matter that is outside the free boundary will descend from the poles to the equator, forming an equatorial zone with particles describing circles around the centre. This would be the start of the formation of a ring. In these calculations, a considerable mass in the centre is necessary at the outset to avoid problems with angular momentum when the condensation is formed. Also, the absence in the beginning of a central mass asks for very massive rings; this calculation is based on Green's integral theorem and the equilibrium condition. One of the conclusions in §19 is that for a mean density ρ and angular velocity ω of a ring, there exists a lower limit of the density for stability:

$$\rho > \frac{\omega^2}{2\pi}.$$

How do we get successive formation of rings? In §22 of the chapter it is argued that after cooling, with as a consequence the rapid formation of an equatorial zone, this formation will entrain extra cooling, making the process discrete. In this process, Bode's law should play a part. This law is an observational rule that asserts that the distance x_n of the nth planet from the Sun is governed by the formula

$$x_n = a + b^n,$$

where a, b are constants and the planets are numbered sequentially, beginning with Mercury 1, Venus 2. The internal dynamics is described by a remark about the velocity distribution of the particles within the equipotential surfaces. There will be angular rotation with many different elliptic orbits. By shocks, the radial velocities will be diminished, making the rotation, according to Roche, more and more uniform. This implies a modification of Laplace's hypothesis, which assumed the inner nebula to be gaseous.

The Assumption of Uniform Rotation

Laplace's model is based on uniform rotation of the nebula with the assumption that the uniformity is caused by frictional effects of the particles in the nebula. Writing

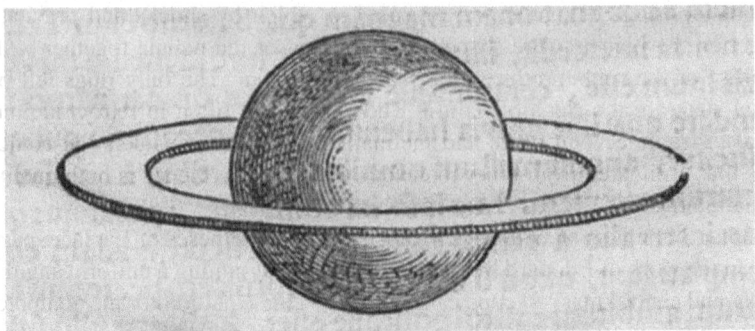

Fig. 11.5 Drawing by Christiaan Huygens in 1656 based on his observations of Saturn. Huygens was the first to argue that the rings are not extensions of Saturn, but consist of debris

down Navier's equation and the continuity equation (§25), we note by a similarity calculation that by analysing an atmosphere, we can immediately obtain results for atmospheres of other sizes. Helmholtz finds that for the atmosphere of the Earth, say of size 8 km, frictional effects would be noticeable on a time scale of 10^{13} years. Poincaré concludes that for an atmosphere the size of the solar system, it would take 10^{22} years, which makes the effect of friction much too weak.

To analyse the consequences, Poincaré considers a thought experiment. One can view the atmosphere as built up of an infinite number of rotating rings, each of them with different angular rotation ω. Taking the friction (or viscosity) to be zero in Navier's equations, we can solve the system under different assumptions of the rotational characteristics. Assume also that the nebula has a very condensed, dominating mass M. The form of the equipotential surfaces depends essentially on the adopted rotation laws of the rings. Assuming, as Laplace and Roche do, that rotation is uniform, we recover the equipotential surfaces found earlier; they lead to the formation of rings by cooling. However, assuming an adiabatic law for the rotational velocities, we obtain for the equipotential surfaces the equation

$$-\frac{\Omega^2}{2y^2} + \frac{M}{\sqrt{x^2 + y^2}} = C,$$

with Ω and C constants. Analysis of this relation shows that no rings can be produced in this case, so the assumption of uniform rotation is essential (§28).

The Stability of the Rings of Saturn

Most studies of rings are concerned with the rings of Saturn, see Figure 11.5. These rings might be solid or fluid or consist of massive fragments revolving around Saturn. However, they cannot be solid, since they are translucent, and they cannot be a continuous medium, a fluid, since the light through the rings is not refracted. This leaves the hypothesis that the rings consist of individual large particles or debris.

In addition, Laplace has shown that a solid homogeneous ring cannot be stable. James Clerk Maxwell considered in 1859 the stability of a ring consisting of individual particles, moving in circles at equidistance. After writing down the equations of motion and linearizing around the solution of constant rotation with angular velocity ω, Maxwell finds stability for a given mass of a ring if the angular velocity ω exceeds a certain limit value.

Maxwell also studied the stability of a fluid ring, finding a limit superior for the density in the case of stability. For Saturn, this upper limit is incompatible with the lower limit for the density that we found in §19. So this is a theoretical argument against the fluidity of the rings of Saturn.

The Stability of the Rings of Laplace

Immediately after the formation of a ring, the density is low, and the fluid particles follow Kepler's laws; the rules for the upper and lower limits of the density are both satisfied. In the beginning, the ring is stable. During the cooling of the ring, the density increases, and the upper limit might be exceeded. Also, uniformization of the rotational velocities by friction destabilizes, so the ring will break up into fragments. At some time, such gaseous parts may collide with each other and merge to produce a globular mass and finally a planet. Tidal forces are expected to synchronize the periods of rotation and of revolution, and the influence of tidal forces will decrease with the distance from the Sun.

The Formation of Satellites

A gaseous mass, supposed to be a protoplanet, can become in turn a planetary nebula, producing a central mass with satellites. Such a planetary nebula will be elongated by tidal forces from the Sun, and it will present always the same points towards the central star. Also, the period of revolution around the central star and the rotation around its axis will become equal (§43). Two possible cases present themselves.

1. The fluid mass is homogeneous. Its density is ρ, and it rotates with angular velocity ω around the x-axis and experiences attraction from its various parts and from the Sun (§§45–48). The Sun is placed far away on the y-axis, i.e., on the equator of the rotating nebula. The nebula experiences forces from its own mass, the centrifugal forces, and the tidal forces of the Sun. This produces a total potential V, which is the sum of these three constitutive potentials. Can we identify the equipotential surfaces of a rotating ellipsoid with the equipotential surfaces of V? The answer is affirmative, with as one of the conditions for stability that the ellipsoid rotate around the smallest axis. For some values of the parameters, one recovers the classical Maclaurin and Jacobi ellipsoids; see Section 11.2. Assuming that the mass

of the planetary nebula is much smaller than the mass of the Sun, one obtains the stability condition

$$\frac{\omega^2}{2\pi\rho} < 0.046.$$

This puts an upper limit on the radius of any satellite system; it is fulfilled for Jupiter's system of satellites.

2. In the case of a strong condensation of the mass of the nebula (§49), the gravitational potential of the nebula is dominated by the central mass. Because of the centrifugal and tidal forces, the total potential will not be spherical symmetric, and the resulting dynamics will, in the course of time, align the nebula in the direction of the Sun. The rotation of the nebula will accelerate, and during this process, the dimensions of the resultant satellite system will change.

The Earth's Moon is exceptional in this scenario. It is relatively large and has an eccentric orbit. According to Roche, it is more probable that the Earth–Moon system was formed as a double planet, but this brings us rather far from the theory of Laplace.

Roche gives a good explanation for the presence of the rings of Saturn (§52). An ellipsoid at the distance of these rings would be destabilized by tidal forces of Jupiter. Satellites cannot be too close.

One can make various objections against the theory of Laplace. One is that the time scales of the various dynamical processes are too long for the time of existence of the solar system. Another objection is that there are satellites in retrograde motion, which is not in agreement with the formation of satellites and planets in a rotating nebula. There are other problems to account for, such as the origin of the Earth–Moon system.

In what follows, Poincaré considers more recent theories. The theories of H. Faye and of R. du Ligondès can be considered a modernization of the theory of Kant.

Faye's Hypothesis (Chapter 4)

Hervé Faye (1814–1902) assumed that the cosmogonic system consisted originally of a chaotic system of material with rotation and turbulent motion. This system was spherical and homogeneous, and contained slowly moving turbulent parts. Ring formation took place, but inside the nebula, in contrast to the theory of Laplace. In the beginning, the rotation is direct; for rings formed after some time, the rotation is retrograde. The implication is, for instance, that the Earth is older than Jupiter and Saturn and even older than the Sun. For Laplace, the comets are intruding alien elements, while for Faye, they are original ingredients of the system. Since the mass of the Sun is increasing slowly, Poincaré shows in §65 that the planets will move nearer to the Sun with time. The implication is that most of the planets were formed much farther away than their present positions.

To explain the near-circular and near-coplanar motion of the planets is not easy with these assumptions.

The Hypothesis of du Ligondès (Chapter 5)

R. du Ligondès assumed that the original cosmogonic system was chaotic or in random motion with small filaments and condensations, but without turbulence and rotation. Interestingly, Poincaré notes that in contrast to the hypothesis of Kant, this assumption is not in conflict with conservation of angular momentum. The reason is that the geometric sum of all the velocities of this chaotic system will be small, but nonzero. This will produce rotation of the system after long evolution (§68). One can actually make a rough estimate of the ratio of the geometric sum of the velocities and the arithmetic sum of the velocities; based on an initial chaotic nebula size of 10^5 AU (1 AU is the average distance between Earth and Sun), and assuming the present amount of angular momentum of the solar system, the ratio will be $1/30\,000$. Collisions between the molecules will lead to loss of energy and therefore contraction of the nebula; the velocity field perpendicular to the equatorial plane being more random than parallel to this plane will lead to loss of energy by collisions in the perpendicular direction and so to flattening of the system. This flattening process increases, because an increase of oscillation frequencies will evoke a decrease of the amplitude (§71); this ties in with the existence of an adiabatic invariant (see comments).

In this theory, the central condensation grows, and the system becomes more and more flattened with internal condensations. One question is whether these internal condensations—protoplanets—will move in near-circular orbits, but a more fundamental question is whether such a system will evolve in agreement with the kinetic theory of gases. According to the hypothesis of du Ligondès, the original nebula evolves from random motion to the special form of the present solar system, while one expects a gas to move to a state of greater disorder. A difference with the kinetic theory of gases is, however, that molecules in a gas are expected to collide without loss of energy, and particles in the nebula may lose heat and may cluster.

For Poincaré, the hypotheses of du Ligondès are an occasion to explore the internal dynamics of gases in §§74–86. Starting with a mechanical system consisting of many material points, he derives the virial theorem:

$$2\bar{T} + \bar{V} = 0,$$

with $\bar{T}$, $\bar{V}$ the time averages of the total kinetic energy and the virial V, given by

$$V = \sum (xX + yY + zZ),$$

with (X, Y, Z) the force acting on a material point at position (x, y, z).

For a gas enclosed in a container, this leads to the relation

$$3pv = 2\bar{T},$$

with p the pressure in the container, v its volume.

In the chaotic nebula of du Ligondès, the interacting force is Newtonian. Introduce

$$W = \sum_{ik} \frac{m_i m_k}{r},$$

with r the distance between the masses m_i and m_k. From the virial theorem, we obtain

$$\bar{W} = -2C,$$

with C a constant if there are elastic collisions only. So in this case there will be no tendency to form a central condensation. If the collisions are not elastic, C will decrease with time, so the distances r have to decrease, and we will have a central condensation. The collisions have to terminate at some time or the condensation will grow indefinitely.

Interesting is the analysis of the density ρ of a liquid and the velocity distribution of the fluid elements. At first, assume an incompressible fluid enclosed in a container. Using the continuity equation in phase space (Liouville equation) and the incompressibility, one finds that ρ does not vary along the trajectory of a fluid element (or molecule). If this trajectory fills the container completely, the density will be constant in the container (§77). If, in addition, the equations of motion admit a first integral J, and a trajectory fills completely the surfaces $J = $ constant in space, then the density will be a function of J: $\rho = \rho(J)$.

This can be extended to the case of an incompressible fluid with n degrees of freedom enclosed in a container. Assuming Hamilton's canonical equations of motion for the dynamics and analogous assumptions on trajectories filling up the phase space, we conclude that the density of the fluid will be constant. If the equations of motion admit k integrals $J_1, J_2, \ldots, J_k$, the density will be a function of these integrals, provided that a trajectory completely fills the $(2n - k)$-dimensional surface

$$J_1 = \text{constant}, \quad J_2 = \text{constant}, \quad \ldots, \quad J_k = \text{constant}.$$

If we have only one first integral, the energy E, we can derive Maxwell's law for the velocity distribution of molecules in a gas.

In the case of the chaotic nebula, the gas is not contained in a vessel, and there are some modifications. First we have extra integrals related to the motion of the centre of gravity. However, taking coordinates with respect to the centre of gravity, Maxwell's law still applies. Elastic collisions and near-collisions do not change the picture; genuine collisions will be rare. Friction will make the rotation more and more uniform, and all the matter will have the same direction of motion.

The theory still has problems with explaining direct versus retrograde motion, the mass differences of the planets, and their ages.

The Capture Hypothesis of See (Chapter 6)

A different proposal came from the American astronomer T. J. J. See (1866–1962), who assumed the presence of the Sun and its original nebula but with the acquisition of planets and moons by the capture of passing bodies. The mechanism would be gravitation in combination with the resistance of a very large nebula around the Sun.

Poincaré considers the influence of a resisting medium on a Keplerian orbit by calculating the changes of angular momentum, eccentricity, and semimajor axis by an averaging method. One of the conclusions is that the eccentricity diminishes. Qualitatively, this can be understood because at perihelion, the velocity and hence the resistance will be optimal; this will also result in a decrease of the semimajor axis. A passing body in a hyperbolic orbit might thus be captured in a highly eccentric orbit, and subsequently, the orbit will become less eccentric, and the distance to the Sun will decrease.

It is interesting to look at this scenario in the context of the restricted circular three-body problem with the incoming body as the third mass point. Poincaré shows that it is possible that under the resisting influence of a surrounding nebula, such a small body can be captured either by the Sun or by Jupiter. See's hypothesis explains the small eccentricities of the planets, but it does not explain the directions of motion and the near-coplanarity.

See wrote a review [See 1912] of Poincaré's book on cosmogony, noting that not all of his work could have been seen by Poincaré, since some of it appeared after the lectures were finished. See adds that he assumes the original nebula to be asymmetrical, consisting of at least two streams that settle in a plane. This would explain the presence of the small inclinations of the planets. Comets and meteoric dust enter the nebula from all directions, building up the larger bodies. According to See, the theory of Laplace and its modifications should be rejected.

Comments on the Cosmogonic Hypotheses

1. In §20 of Chapter 3, we find a similarity calculation for the equipotential surfaces.
2. The formation of rings as a discrete process in Chapter 3, §22, is rather qualitative. It needs to be elaborated quantitatively.
3. The term "planetary nebula" is nowadays used exclusively for a star with surrounding nebula, but we retain it, following Poincaré, also for a protoplanet with surrounding nebula (§43). A weak point here is the lack of estimates of time scales for tidal forces to become effective.
4. Roche's analysis, presented in §52, regarding the breaking up of satellites that are too close to the central body is still in use. The lower limit for a stable distance is called the Roche limit.
5. To analyse some of Faye's assumptions regarding the evolution of the solar system, Poincaré considers in §65 adiabatic approximations using averaging. See [Verhulst 2000] for a modern treatment. In §71, a similar treatment is concerned with the relation between frequency α and amplitude x_0 of oscillations in a

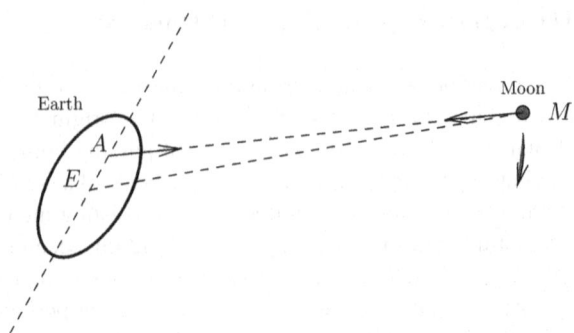

Fig. 11.6 The Earth, covered by oceans, is tidally deformed by the Moon; the resultant force does not act along the line E–M but along A–M

slowly flattening nebula. This leads to the (now) well-known adiabatic invariant $x_0^2 \alpha =$ constant.

6. The virial theorem was proved by Poincaré in §74; applications are given in §§75–76.

7. The analysis of the density of a fluid in a container (§77) supposes that trajectories fill up the container completely. This is a slightly stronger requirement than the result of the recurrence theorem, and it anticipates ergodicity.

8. The mechanism proposed by See seems to be a mixture of quantitative and qualitative arguments, as in fact most cosmogonic hypotheses are. This will be discussed again at the end of the section.

Tidal Evolution According to Darwin

Chapter 7 contains a description of the tidal theory of George H. Darwin (1845–1912), son of the famous evolutionary biologist Charles Darwin. It traces the consequences of tidal friction for the problem of a body covered by a liquid and accompanied by a satellite. In addition, some aspects of tidal friction in the three-body problem are discussed.

In a simplified view of the tides, the Earth is completely covered by the oceans, and the Moon is represented by a point mass. The oceans take the form of an ellipsoid, but its major axis will not be aligned with the axis joining the Earth's and Moon's centres of gravity; see Figure 11.6. This nonalignment is caused by the rotation of the Earth about its axis, the motion of the Moon around the Earth, and the retardation of the moving oceans caused by friction. The implication is that the resultant force is not central and has a nonzero moment with respect to the line $E - M$; see again Figure 11.6. This slows down the rotation of the Earth about its axis; it acts on the Moon as a tangentially propelling force.

A global way of looking at tidal evolution gives additional insight. The Earth–Moon system undergoes friction, whatever its specific form, while the total angular momentum of the system is constant. The friction will reduce the speed of rotation of the Earth, and so the orbital motion of the Moon has to increase, resulting in an increase of the distance Earth–Moon.

According to Darwin, primeval Earth was entirely liquid, so that tidal friction was originally much more effective. Following Darwin, Poincaré assumes first that the lunar orbit is in the plane of the equator of the Earth and that the eccentricity is zero. Because of symmetry, motion that starts in the equatorial plane will remain in this plane. Will the orbit remain circular? From a calculation of the rotational moment and the kinetic energy of the system, it can be shown that this is indeed the case.

There are already interesting consequences. Introduce the variable y for the rotational velocity of the Earth around its axis and the variable

$$x = \frac{1}{\Omega^{1/3}},$$

with Ω the angular velocity of the Moon around the Earth. The total angular momentum h being constant is expressed by the equation

$$x + y = h.$$

From the expression for the total energy, we find with the angular momentum integral that for stationary values of the energy, we have

$$x^3 y = 1.$$

According to Darwin (§99), equilibria will be found at the intersections of two curves; see Figure 11.7. If the curves do not intersect, there exists no equilibrium. During its tidal evolution, the system will remain on the axis $x + y = h$. For a system like Earth–Moon, the month is much longer than the day; its position in Figure 11.7 will be between the points C and D. The evolution will be towards point D, the equilibrium where the length of day and month have become equal. During this evolution, the Moon will be more and more removed from the Earth. Comparable reasoning applies to other satellites in the solar system.

Other effects to be considered are the influence of the Sun on the evolution of the Earth–Moon system and the shrinking of the Earth because of cooling. The scenario sketched here was for Darwin a reason to conjecture that the Moon was formed by a breakup of part of the Earth.

More General Analysis

It is important to analyse the evolution by tidal friction, starting with more general initial conditions, i.e., inclination and eccentricity nonzero. This technically much more complicated scenario is partly based on the tidal theory in Poincaré's lecture notes [Poincaré 1890a, nr. 13, Vol. 3]. After a detailed and long analysis (§§103–

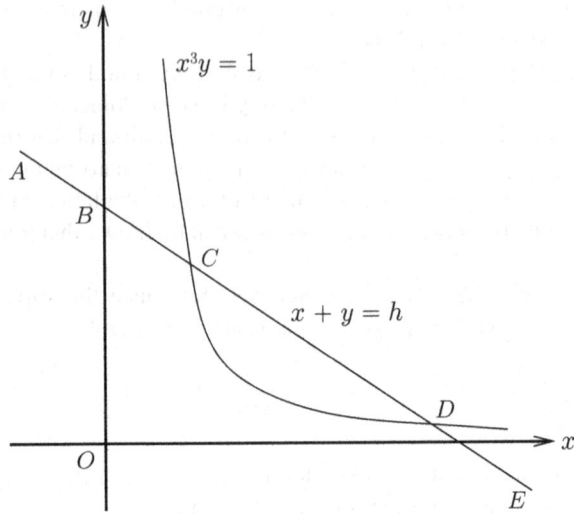

Fig. 11.7 Tidal evolution according to Darwin. The two-body system is located during its evolution on the straight line $x + y = h$ and will move by tidal forces. An unstable equilibrium is located at C, a stable one at D. On the segment AB there is retrograde motion, on BC direct motion; starting on the segment AC, the satellite will plummet into the planet. On the segment CD, on which the Earth–Moon system is located, the evolution is towards a stable equilibrium D in direct motion, while the month is longer than the day. On the segment DE the evolution is towards stable equilibrium D in retrograde motion, while the month is shorter than the day. In D, month and day are equal, and the system is in 1 : 1 resonance

122), an interesting conclusion is that one can draw a similar diagram as in Figure 11.7. Stable solutions with eccentricity and inclination zero exist and can be reached by evolution, depending on the physical parameters and initial conditions.

One can estimate the influence of the Sun on the tidal evolution of the Earth–Moon system; this turns out to be insignificant. On the other hand, cooling of a protoplanet involves contraction and so an increase of the rotational speed. This may have been important, especially for the larger, outer planets.

The origin of the Moon remains an open problem. As indicated above, Darwin conjectured that at some point, a quantity of mass was removed from Earth, probably by a solar tidal resonance, which was possible, since the Earth was still liquid. This mass would have drifted away from the Earth because of tidal friction and become the Moon.

Another possibility mentioned by Poincaré is fission of a rotating protoplanet as described by him in his monograph on rotating fluid bodies; see Section 11.2.

Comments

1. Darwin's conjecture on the origin of the Moon seems to be in contradiction with the theory of Roche. A satellite that is too close to a planet or star will be destroyed by tidal forces, producing large fragments or a ring of debris. As noted earlier, the boundary for this distance limit is called the Roche limit. On the other hand, one could conjecture the violent ejection of a quantity of material that subsequently, outside the Roche limit, agglomerated to become the Moon.
2. The qualitative reasoning of Darwin regarding the outcome of tidal evolution reproduced in §99 and illustrated in Figure 11.7 was discovered again in 1973; see [Counselman 1973]. The analysis by Poincaré of tidal evolution from general initial conditions was an improvement over Darwin's results.

Astrophysical Considerations

Chapter 8 discusses the origin of the heat of the Sun and the Earth. Apart from mechanical arguments, this also involves gas dynamics and thermodynamics. The Sun loses an enormous amount of heat, and sources of chemical origin, meteoric collisions, and heating by contraction of a gas have been proposed. All of them lead to short time scales for the existence of the Sun, millions of years, and this makes them unsatisfactory. In fact, on a much smaller scale, this also applies to the heat production of the Earth. According to Poincaré, the solution has to be found in the radioactivity of elements such as radium and uranium and probably other, still unknown, energy sources.

As an illustration, without much relevance for the Sun, Poincaré considers the question whether a hot radiating sphere consisting of a perfect gas can maintain its temperature distribution by contraction. It turns out that this is possible for a gas consisting of atoms and during a computable interval of time; the formation of molecules during contraction complicates the calculations.

In Chapter 11, Poincaré describes the theory propounded by the physical chemist Svante Arrhenius (1859–1927). An essential ingredient of this theory is the radiation pressure in space. Following Maxwell's electromagnetic theory of light, one can explain radiation pressure by the pressure and tension of the electric and magnetic field lines. It is also possible to give a purely thermodynamic account. Poincaré proposes in §180 as a thought experiment a vessel without matter but filled with radiation; it has black walls and a piston closing the vessel without friction. The vessel is kept at a constant temperature. Using the assumption of thermodynamic equilibrium and reversibility, one finds that the radiation pressure has to be proportional to the total radiation energy per unit volume.

A particle in a neighbourhood of the Sun will be attracted by gravitation (proportional to its mass) and repelled by the radiation force (proportional to its area directed towards the Sun). In this way, radiation pressure acts on the tails of comets when they are close to the Sun. Dust particles are pushed far away, where they can be observed as a nebula. In the scenario of Arrhenius, the Sun cools at the outside, while the inside remains hot. When stars are nearly extinguished, they may collide, producing a nova, a new star. Such a collision produces two material jet streams with a certain angular momentum. What follows is that in the nebula that emerges, a new star is formed, also helped along by accretion of meteorites. Arrhenius assumes that the universe is infinite and that this process will go on without end. This seems in contradiction with the second law of thermodynamics (the Carnot–Clausius principle), but according to Arrhenius, the second law of thermodynamics applies to stars like the Sun but not to nebulae. This makes possible a permanent reorganization of the components of the universe. Poincaré compares this process to the activity of what is known as Maxwell's demon. In Maxwell's thought experiment, one considers two containers, A and B, filled with gas and connected by a tube. All collisions of the molecules are perfectly elastic. A "demon" places itself in the tube and elastically repels all molecules with velocity below a certain threshold moving in the tube in the direction from A to B; all molecules with a higher velocity are allowed to pass through. The demon admits all slow molecules from B that want to move to A and repels the fast molecules back into B. No energy will be lost, and eventually, all the slow molecules will be in container A, the fast ones in B.

Arrhenius's conjectures are rather bold, but Poincaré spends many pages in analysing them and their consequences. The difficulty is that the conjectures are mostly qualitative in nature and lacking in quantitative details, so to invalidate the theory is possible only using first principles. After analysing the behaviour of gases under various conditions, Poincaré concludes that the question has not been settled definitively, but that he expects that the second law of thermodynamics will remain valid in the Arrhenius scenario, but with the disorder delayed.

A confusing element in the discussions around 1900 was that a number of scientists wanted to account for the origin of both the solar system and the spiral nebulae. There were not yet enough data on the differences between nebulae and galaxies. One interesting remark by Poincaré in §199 is concerned with the velocity distribution of stars in galaxies. Recognizing that galaxies consist of swarms of stars, Poincaré doubts whether these galaxies are in an equilibrium state. His argument is that smaller bodies in a Maxwell distribution of velocities should have a much higher velocity than the massive stars. Comets, for instance, move in general faster than stars, but not fast enough to fit in a Maxwell distribution.

Discussion

Until around 1900, the nebular hypothesis was the dominant theory, but there was already criticism. F. R. Moulton (1872–1952) and T. C. Chamberlin (1843–1928) objected to the Laplacian theory of successively separating rings in a nebula followed by condensation to protoplanets. Moulton considered the possibility of a close encounter of two stars, one of them the Sun, that resulted in drawing filaments of matter out of them. This matter could go on forming planetary objects around the Sun. A second hypothesis of Moulton and Chamberlin was that matter would condense into small solid particles, the "planetesimals." The solid particles were expected to accrete to larger bodies.

These Chamberlin–Moulton theories were not complete as a mathematical–physical theory; they provoked discussion and caused many controversies, but their acceptance remained mostly restricted to the United States. The French astronomers followed Poincaré, who ignored Chamberlin and Moulton. The British theoretical (astro)physicist and mathematician James Jeans (1877–1946) analysed the fission theory—see Section 11.2—and the influence of tidal forces. He was joined in these discussions by Harold Jeffreys (1891–1989), who showed that collisions of interplanetary matter produced a hot gas instead of condensations. Other astrophysicists pointed out that stellar material at very high temperatures would dissipate into space. This causes problems in viewing the solar system as originating from a near-collision between stars or, alternatively, as a "failed double star."

In 1913, Karl Schwarzschild (1873–1916), one of the most prominent astronomers of that time, wrote a review [Schwarzschild 1913] of Poincaré's "Hypothèses cosmogoniques." He notes that the Chamberlin–Moulton theory is not discussed in the book, but he does not greatly object, since he prefers the updated Kantian hypothesis, which does not—in contrast to the hypothesis of Laplace—ask for perfect regularity, while the laws of mechanics are saved by the ideas of See regarding a condensation in the centre and a resisting medium outside. Schwarzschild also notes that a number of astronomers compare spiral nebulae wrongly with the solar system; he estimates that the size and mass content of a spiral nebula are orders of magnitude greater than those of the solar system.

Schwarschild also discusses the theory of Arrhenius, ascribing to him "fascinating imaginative power." He agrees with Poincaré that at most a delay of the transition to a disordered state can be assumed by Arrhenius's cyclic process. On the other hand, he does not exclude the possibility that on a very long time scale, say 10^{50} years, a cyclic process in the universe takes place.

In fact, as astrophysics rapidly developed in the twentieth century, more and more theoretical insight and more quantitative results had to be accounted for. A striking example is the theory of H. O. G. Alfvén (1908–1995), a plasma physicist. Alfvén showed that a hot rotating Sun possessing a magnetic field will reduce the speed of rotation of the Sun. For extensive discussions and references, see [Brush 1996].

The solar system is a very complex physical system, and at present, there is still no satisfactory theory to account for its origin.

Chapter 12
Poincaré's Address to the Society for Moral Education

Nowadays, scientists are not too fond of "moral societies." Moralizing as such tends to arouse suspicion, for it often takes place for reasons that are not mentioned explicitly. Also, societies with a moralizing agenda are often sponsored by wealthy business executives, a class with which many intellectuals do not want to be associated.

Poincaré addressed the Ligue française d'éducation morale on June 26, 1912, just three weeks before his death on July 17. The tremendous problems in French society that had been uncovered by the Dreyfus affair (see Section 4.8) provided one of the reasons to discuss moral problems in the nation as a whole. Another, related, aspect was the emergence of the socialist movement and the drive for the emancipation of workers. At that time, the churches and most religious leaders had announced themselves firmly on the side of the establishment, thereby estranging many active and unemployed workers from religion.

The international atmosphere was very strained, and indeed two years later, the First World War would engulf Europe. It is remarkable that also on moral education, as he had on other, more scientifically oriented, topics, Poincaré touched an essential point: society is made up of a very diverse collection of individuals, and it is important that they respect one another and cooperate. Above all, they must *avoid hatred* between different social groups. Indeed, such a message remains valid for all times.

It is of interest that Poincaré, in a side remark, also connected "loss of morals" with "loss of beauty." That is, he links ethics with aesthetics. Perhaps this remark is typical for a mathematician, for whom mathematical truth is intimately linked with the aesthetic qualities of the mathematical structures and arguments being created. We follow the text as presented in [Appell 1925a].

Ladies and Gentlemen: Today's meeting assembles people with very different opinions who approach each other only because of common good intentions and the same desire for well-being. Nevertheless, I do not doubt that they will understand each other easily, for although they have different opinions about the means, they agree about the goal they want to reach. And it is only this that counts.

F. Verhulst, *Henri Poincaré: Impatient Genius*, DOI 10.1007/978-1-4614-2407-9_12,
© Springer Science+Business Media New York 2012

Recently, one could read about a conflict of morals; one can still read it on the walls of Paris, where a meeting with discussion is announced. Does this conflict exist, should it exist? No. Morals can be established for many reasons; some of these are transcendental, and perhaps these are the best ones and certainly the noblest, but these are also the ones on which opinion differs. There is at least one argument, perhaps a bit down to earth, on which we cannot disagree.

In reality, man's life is a permanent struggle. The forces directed against him are doubtless blind, but they are so formidable that they quickly floor him, they allow him to perish, they flood him with a thousand sorrows if he is not always prepared to withstand them. If, sometimes, we enjoy relative quiet, this is because our ancestors have struggled mightily. If our exertions, our vigilance, were to weaken for a moment, we would lose all the fruits of their struggle, everything they have gained for us. So, humanity constitutes an army in the midst of war.

However, every army needs discipline; it is not enough if one submits to discipline only on the day of battle. The army has to submit to it even in times of peace. Without discipline, defeat is certain, and no amount of bravado can prevent it.

What I have said just now applies as well to the struggle humanity has to maintain throughout our lives: the discipline she has to accept is called "morals." On the day she forgets this, she will be brought down by an advance, and she will be thrown into an abyss of evil. Moreover, she will undergo on this day a deterioration; she will find herself less beautiful, and, let us say, diminished. We would be distressed not only because of the bad things that would follow, but also because of the obscuration of beauty.

About all these points we think alike; we all know where we have to go. Why would we disagree about knowing which road to walk? If reasoning could accomplish something, agreement would be easy. Mathematicians never disagree about how to prove a theorem, but this involves something entirely different. To formulate morals by reasoning is useless; in these affairs there is no reasoning without counterarguments.

Explain to the soldier how many bad things will proceed from defeat and that it will even touch upon his personal safety. He could always answer that his safety would be better guaranteed if the others would fight. If the soldier does not answer in this way, it is because he is muted by some force that silences all reasoning. What we need are forces like that. Now, the human soul is an inexhaustible reservoir of forces, a fertile source, a rich source of energy of motion. Emotional feelings constitute this energy of motion, and moralists should, so to speak, capture these forces and point them in the right direction, just as engineers tame the forces of nature and mold them to industrial requirements.

But at this point diversity arises: to run the same machine, engineers can arbitrarily appeal to steam or to hydraulic power. In the same way, teachers of morals can, according to their taste, put some psychological force into motion. Of course, everybody will use the force that he experiences in himself. He could handle outside forces or those that he could borrow from a neighbour only awkwardly. They would be lifeless and without effect in his hands; he would give up, and he would be right.

Because their weapons are different, their methods have to be, and why would they take exception because of this? And still, the morals one teaches are always the same. If you visualize general well-being, if you appeal to compassion or to the sentiment of human dignity, you always end up with the same rules. You cannot forget those, or the nations will perish, while at the same time suffering will multiply and humanity begin to decline.

If all those people are fighting the same enemy with different weapons, why do they recall so rarely that they are allied? Why do some enjoy the failure of the others? Are they forgetting that each of these failures is a triumph for the eternal enemy, a diminishment of the common inheritance? No, we need all our forces too much to have the right to neglect one; also, if we exclude no one, we banish hate.

Certainly, hate is also a force, a very powerful force. But we cannot use it, for it belittles, because it is like eyeglasses that provide only a rough outline. Even between nations, hate is disastrous, and it does not produce true heroes. I do not understand how across certain frontiers, one believes that it is possible to profit from being patriotic with hate. This is contradictory to the instincts of our race and its traditions. The French armies have always battled for someone or something and not against someone; and they have not fought the less for that.

If, within our frontiers, the parties forget the great ideas that were honourable and were the reason for their existence, if they remember only their hate, if one says, "I am against this," and another answers, "I am against that," the horizon shrinks, as if the clouds had descended and veiled the summits. The vilest means are used, one does not stop at calumny, nor at denunciation, and those who are amazed at this become suspect. One sees the emergence of people who seem to use their intelligence only to lie and their hearts for betrayal. One observes souls who have no bad taste, but who, as soon as they take refuge under the same flag, can endorse many things and even show admiration. Confronted with so much opposing hate, one hesitates to wish the failure of one that will be the triumph of the others.

This is what hate can achieve, and this is exactly what we do not want. So let us approach each other, get to know each other, and in this way respect each other, so that we can strive for the ideal community. Guard against the imposition of uniform means, for such cannot be realized. Moreover, it is undesirable: uniformity means death, because it is a closed door to all progress. Also, every constraint is sterile and detestable.

People vary; some are rebels, concerned with a single maxim that leaves everyone else indifferent. I cannot know whether you are going to announce this decisive maxim, but I would forbid you to pronounce it! ...But you will see the danger. These who have not received a similar education are bound to hurt themselves in life. Their souls are going to be shaken by repeated shocks; they will change, perhaps they will find another faith. What will happen if the new ideas they are going to adopt are those that their old masters presented them precisely as the denial of morals? Will this frame of mind vanish in a day? At the same time, their new friends will teach them not only to reject what they have worshipped, but to despise it. They will not keep tenderly the memory of the rich ideas that have filled their souls and that have survived their faith. In this general ruin, their moral ideal risks being swept away; too old to undergo a new education, they lose the fruits of the old one!

This danger can be exorcised, or at least diminished, if we learn to speak only respectfully about all the sincere efforts other people near us are making. This respect would come more easily to us if we knew ourselves better.

And this is exactly the purpose of the "Ligue d'éducation morale." Today's celebration, the lectures you are going to hear, will be sufficient proof to you that it is possible to have a fervent belief and to do justice to the beliefs of someone else, and that finally, in various uniforms, we are only several divisions of the same army, battling side by side.

Chapter 13
Historical Data and Biographical Details

We summarize here some data and events that are characteristic of the social and political climate in which the Poincaré family lived:

- The Duchy of Lorraine was incorporated into France in 1766. Part of Alsace, the Republic of Mulhouse, became a region of France in 1798.
- 1815: Final defeat of Napoleon and his abdication as emperor. Beginning of the restoration of reactionary powers (Ancien Régime).
- 1815–1848: Reign of the kings Louis XVIII, Charles X, and Louis-Philippe.
- 1828–1892: Léon Poincaré, father of Henri.
- 1830–1897: Eugénie Launois, Henri's mother.
- 1854–1912: Henri Poincaré.
- 1856–1919: Aline Poincaré, Henri's sister.
- 1860–1934: Raymond Poincaré, Henri's cousin and president of the republic (1913–1920).
- 1848–1851: Second Republic.
- 1852–1870: Reign of Napoleon III, Second Empire.
- 1870–1871: Franco-Prussian War, followed by the loss of Alsace-Lorraine; Nancy remained French.
- 1871–1875: Political confusion after the defeat and loss of Alsace-Lorraine. Continuing fights between republicans and royalists.
- 1875: Failure of the royalist restoration and foundation of the Third Republic.
- 1905: Following the political upheavals of the Dreyfus affair, diplomatic relations between France and the Vatican are broken, followed by secularization of all public affairs.
- 1914–1918: First World War with main battlefields in France and Belgium.

F. Verhulst, *Henri Poincaré: Impatient Genius*, DOI 10.1007/978-1-4614-2407-9_13, © Springer Science+Business Media New York 2012

Fig. 13.1 Vladimir Arnold,
mathematician

Biographical Details

Newtonian mechanics, in particular celestial mechanics, was a standard part of the university curriculum of mathematics and physics until 1940. So it is not surprising that many great names in science are attached to this field. In the following biographical sketches, we restrict ourselves to the scientists mentioned in relation to Poincaré's life and work.

Max Abraham (1875–1922). German-Jewish physicist who completed his doctorate under the direction of Max Planck. He showed in 1900 by experiments the relation between electromagnetic inertia and velocity; working in Göttingen and outside Germany, nearly all his work was on electromagnetic theory. He was a strong opponent of relativity.

Hannes O.G. Alfvén (1908–1995), Swedish electrical engineer and plasma physicist. He was educated in Uppsala and Stockholm, became a professor in Stockholm, and moved in 1967 to the USA. He received the Nobel Prize in physics in 1970, but he often had trouble in publishing his papers because of clashes with traditional opinion in physics.

Paul Appell (1855–1930); see Section 4.3 and Figure 4.3.

Vladimir I. Arnold (1937–2010), student of Kolmogorov, was a professor in Moscow who obtained fundamental results in Hamiltonian systems, symplectic geometry, and singularity theory. He was a critic of too much abstractness in the presentation of mathematics. See Figure 13.1.

Svante A. Arrhenius (1859–1927), born in Uppsala, Sweden, where he studied and obtained his doctorate (1884). From 1891 he worked (mainly) in Stockholm on physical chemistry. He was awarded the Nobel Prize in chemistry in 1903.

Louis Bachelier (1873–1946); see Section 4.5.

René Baire (1874–1932), French mathematician, educated at the École Normale Supérieure. He attended the lectures of Poincaré in 1894 and edited one of the volumes of lecture notes. He received his doctorate at the Sorbonne in 1899. He became famous for his classification (categories) of functions. He worked at various places, including Bar-le-Duc, Montpellier, Dijon, Paris; his best-known student was A. Denjoy.

Joseph Bertrand (1822–1900), French mathematician. From the age of nine, he was educated by Jean Duhamel and his wife. He attended lectures at the École Polytechnique from the age of eleven. He began teaching at a lycée in 1841, and from 1844 held academic positions and became important in Parisian cultural life. His work on probability strongly influenced Poincaré in his *Calcul des probabilités*.

George D. Birkhoff (1884–1944), was born in Overisel, Michigan, USA, the son of Dutch immigrants. He was educated at Chicago and at Harvard. He became known especially for his work on dynamics and ergodic theory. As a professor at Harvard he was the leading American mathematician of his time, with prominent students R.E. Langer and M. Stone. One of the most important influences on his development was the ideas and results of Henri Poincaré. In 1913 he proved the Poincaré–Birkhoff theorem on periodic solutions; this theorem was formulated by Poincaré in 1912, shortly before his death.

Karl P.T. Bohlin (1860–1939) was born in Stockholm, studied in Uppsala, and became an astronomer, first in Uppsala, later in Stockholm and Lund. The focus of his work was on the motions of planets and satellites in the solar system. He became a professor in 1897.

Ludwig Boltzmann (1844–1906) was born in Vienna, Austria. He studied physics and was a student of J. Stefan. He was professor in Graz, Munich, and Vienna and developed the kinetic theory of gases. At the end of the nineteenth century, there arose controversies and bitter debates about his atomic theory in which prominent scientists took part. Boltzmann was subject to sudden bouts of depression; he committed suicide.

János Bolyai (1802–1860), Hungarian mathematician. He studied at the Academy of Engineering in Vienna and spent 11 years in military service. He developed the basic ideas of hyperbolic geometry by replacing the parallel postulate of Euclidean geometry; unfortunately, he was mostly isolated in his mathematical activity, though the great Gauss recognized the significance of his work.

Pierre-Ossian Bonnet (1819–1892), French mathematician, studied at the École Polytechnique and later took up civil engineering. He returned to the École

Polytechnique in 1844 and became its director of studies in 1871. He was a member of the Académie des Sciences and succeeded Le Verrier in 1878 in the chair of physical astronomy at the Sorbonne. One of his achievements is the Gauss–Bonnet theorem in geometry.

Émile Borel (1871–1956), French mathematician who entered the École Normale Supérieure in 1889. By 1892 he had published six mathematical papers, and in 1893 he received his doctorate. After three years in Lille he returned to the École Normale Supérieure and held subsequently a number of different professorships. He was a versatile mathematician with fundamental contributions in analysis, geometry, and probability.

Jean-Claude Bouquet (1819–1885), French mathematician, educated in Paris at the École Normale Supérieure, doctorate at the Sorbonne in 1851, after which he worked at several places, returning to Paris in 1864, where he lectured at the École Polytechnique and the École Normale Supérieure. His achievements in differential geometry and ODEs were partly done in collaboration with his friend Briot.

Émile Boutroux (1845–1921), French philosopher who married Aline Poincaré. In 1865, he entered the École Normale Supérieure (returning there in 1877), studied in Heidelberg, and received his doctorate in 1874. He defended the position that religion and science are compatible, and one of his students was Henri Bergson. He was elected to the Académie Française in 1912.

Pierre Boutroux (1880–1922), son of Émile Boutroux and Aline Poincaré. Educated at the École Normale Supérieure, he became a mathematician and historian of science. He held positions in various places, including one year in Princeton; he gave up that position at the outbreak of World War I to enlist in the French army. His books on the history of science are still of importance.

Charles Auguste Briot (1817–1882), French mathematician, studied at the École Normale Supérieure and completed his doctorate in 1842. In 1864, he became professor at the Sorbonne and the École Normale Supérieure. His work on elliptic functions, mathematical physics, and ODEs was partly in collaboration with his friend Bouquet.

Ernst H. Bruns (1848–1919), German mathematician and astronomer at the observatory of Leipzig. He proved an important result on the nonintegrability of the three-body problem. A famous student of Bruns was Felix Hausdorff.

Georg F.L.P. Cantor (1845–1918), mathematician, born in St. Petersburg of a Danish father and Russian mother, moved to Germany in 1856, was educated in Zurich and Berlin. Held a chair at the University of Halle and obtained fundamental results in number theory and set theory, strongly opposed by Kronecker. He had a tragic history of mental illness.

Augustin Louis Cauchy (1789–1857), French mathematician who, on the advice of Lagrange, entered the École Polytechnique in 1805. In 1807, he continued with

Fig. 13.2 George H. Darwin, mathematical physicist

civil engineering, which led to work as a constructive engineer. His interest and high-quality productivity in mathematics brought him to the École Polytechnique in 1815. Political and religious problems (he was to a degree a religious zealot) took him to Turin in 1832. His temper and rigid ideas caused him many problems with colleagues.

Thomas C. Chamberlin (1843–1928), born in Mattoon, Illinois, USA. He became a geologist and studied the glacial stages of the Earth, in particular of North America. In 1892, he moved to the university of Chicago, where he developed, together with F.R. Moulton, the planetesimal hypothesis for the origin of the solar system.

Michel Chasles (1793–1880), French mathematician who entered the École Polytechnique in 1812 (doctorate under the supervision of Poisson) and returned there as a professor in 1841. He obtained a chair at the Sorbonne in 1846. His main contributions are in geometry and the history of science. His most famous student was Gaston Darboux.

Gaston Darboux (1842–1917) was born in Nîmes, France. He was appointed to the École Normal Supérieure in 1872, where he lectured till 1881. In 1880, he succeeded Chasles in the chair of geometry at the Sorbonne. His mathematical results are wide-ranging in geometry and analysis, a combination in the spirit of Chasles. He was considered an exceptionally good teacher and administrator. See Figure 4.2.

George H. Darwin (1845–1912), British mathematical physicist (Cambridge), who worked on tidal forces and tidal prediction, also on the fission theory of the Moon in the framework of rotating liquid masses. He was appointed to a chair in astronomy at Cambridge in 1883. See Figure 13.2.

Julius Wilhelm Richard Dedekind (1831–1916), German mathematician, studied in Göttingen under the supervision of Gauss; doctorate in 1852, appointment as lecturer in Göttingen in 1855, cooperation with Dirichlet. He obtained a chair in Braunschweig in 1862; his main contributions are in number theory.

Charles-Eugène Delaunay (1816–1872), French mathematician, engineer, and astronomer. He was educated at the École Polytechnique and the École des Mines. He published technical treatises and, influenced by the books of Laplace, studied methods of celestial mechanics. A characteristic title of one of his books is *Traité de mécanique rationelle*.

J.P.G. Lejeune Dirichlet (1805–1859), German mathematician with parents from Liège, Belgium. After studies in Cologne, he studied in Paris at the Sorbonne, where he obtained striking results in number theory. From 1828 to 1855 he lectured in Berlin (famous student, Leopold Kronecker), from 1855 in Göttingen. He also achieved important results in mathematical physics, for instance in Fourier theory.

Jules J. Drach (1871–1949), French mathematician, born in Alsace; as in the case of Appell, his (farming) family left Alsace in 1871. Although in poor circumstances, he studied at the École Normale Supérieure in 1889, receiving his doctorate in 1898. He coedited lectures of Poincaré. After various positions elsewhere, from 1913 he held a chair at the Sorbonne. Most of his mathematics is on the relations between geometry and differential equations.

Alfred Dreyfus (1859–1935), attended the École Polytechnique; see Section 4.8.

Jean-Marie Duhamel (1797–1872), born in Saint-Malo, France, studied law and mathematics, lectured at the École Polytechnique from 1830 and wrote a popular *Cours d'analyse*. Most of his results are on partial differential equations and include the famous Duhamel principle.

Hervé Faye (1814–1902), French astronomer, educated at the École Polytechnique. His papers are on celestial mechanics and cosmogony. He also had an administrative and political career.

Camille Flammarion (1842–1925), French astronomer who did little research but wrote many popular books, ranging from astronomy to spiritualism, published by his brother Ernest.

Ernest Flammarion (1846–1936), publisher of fiction and popular nonfiction, including the philosophical books of Poincaré. In 2000, the company was acquired by the Italian firm RCS MediaGroup.

Erik Ivar Fredholm (1866–1927), Swedish mathematician, born in Stockholm, studied in Uppsala, and completed his doctorate under the direction of Mittag-Leffler. He became a professor in Stockholm. Inspired by mathematical physics, he did fundamental work on spectral theory, integral equations, and operator theory. The year 1899 represented an important period during which he worked in Paris with Poincaré and other French mathematicians. Hilbert extended a number of his ideas.

Fig. 13.3 Hugo Gyldén, astronomer

Gottlob Frege (1848–1925), German logician and mathematical philosopher. Studied in Jena and Göttingen; influenced, for instance, Peano and Russell.

Lazarus I. Fuchs (1833–1902), German mathematician, studied in Berlin, doctorate under Weierstrass. Worked at the universities of Heidelberg, Berlin, and Göttingen. Famous students: Schur and Zermelo.

Hugo Gyldén (1841–1896) was born in Finland, where his father was a professor of Greek at the University of Helsinki. He became a student of the astronomer P.A. Hansen in Gotha. In Stockholm, he held the position of director of the observatory from 1871 and was an influential astronomer who stimulated celestial mechanics research in Sweden. Famous student: Backlund. See Figure 13.3.

Jacques Hadamard (1865–1963), French mathematician, educated at the École Normale Supérieure. Received his doctorate in 1892 on complex functions. After a few years in Bordeaux, he returned to Paris, and later was appointed to a chair with strong support of Poincaré. In 1896, he gave one of the first proofs of the prime number theorem. He wrote many papers and books; famous students include Maurice Fréchet and André Weil. He fled France in 1940, returned in 1945.

Fig. 13.4 George Henri
Halphen, mathematician

George Henri Halphen (1844–1889), French mathematician, was educated at the École Polytechnique, starting in 1862. He performed military service with great distinction during the Franco-Prussian war of 1870–1871. His doctorate in 1878 was on integral invariants. In 1884, he began lecturing at the École Polytechnique. He was considered brilliant, but he died young, at 44, and his geometrical work is no longer fashionable. See Figure 13.4.

Charles Hermite (1822–1901), French mathematician; entered the École Polytechnique in 1842 but left after a year because of physical problems. He returned as a lecturer in 1848, and was appointed to a chair at the Sorbonne in 1869. His mathematical work is on number theory (he gave the first proof of the transcendence of *e*) and analysis. See Figure 5.1.

David Hilbert (1862–1943), German mathematician, educated in Königsberg (now Kaliningrad), where he became a member of the university staff. Klein arranged his move to Göttingen in 1895. He published on invariant theory, geometry, algebraic number theory, and functional analysis.

George W. Hill (1838–1914), American mathematical astronomer, worked at the Nautical Almanac Office, outside academia. He worked in celestial mechanics and differential equations with periodic coefficients and was in his time the leading American astronomer.

Carl G.J. Jacobi (1804–1851), German-Jewish mathematician, studied in Berlin and was professor at the University of Königsberg (now Kaliningrad). His mathematics was on elliptic functions, theta functions, and on the foundations of mechanics, called Hamilton–Jacobi theory. See Figure 9.1.

Walter Kaufmann (1871–1947), German physicist who obtained the first experimental proof of the dependence of mass on velocity. He supported Max Abraham in the controversies on relativity.

Lord Kelvin ; see William Thomson.

Felix Klein (1849–1925), German mathematician, completed his doctorate under Plücker. He worked on geometry and group theory, non-Euclidean geometry, and complex function theory. He became a professor when he was 23, formulated the Erlangen programme, moved in 1886 to Göttingen. His cultural and mathematical influence was enormous. He made Göttingen a leading centre of mathematics. See Figure 3.6.

Andrei N. Kolmogorov (1903–1987) was an outstanding Russian mathematician. As an undergraduate at Moscow State University, he published eight papers. His activity was wide-ranging, with fundamental contributions in probability theory, functional analysis, and many other areas. In a visionary lecture in 1954 at the International Congress of Mathematicians in Amsterdam, he formulated the celebrated KAM theorem of Hamiltonian dynamics.

Sonya Kovalevskaya (1850–1891), born in Moscow, studied mathematics in Berlin with Weierstrass. Doctorate in Göttingen on recommendation of Weierstrass. She was appointed to a chair in Stockholm in 1884. Her publications are on analysis and mathematical physics. She also wrote two novels.

Leopold Kronecker (1823–1891), German mathematician who was a student of Kummer. He completed his doctorate in 1845 (Berlin), but after that, until 1855, he went into business. He returned to Berlin and was appointed to a chair in 1883. His publications are in number theory, logic, and analysis. His relations with colleagues could be problematic, for instance with Cantor and Weierstrass. See Figure 13.5.

Joseph L. Lagrange (1736–1813), Italian–French mathematician who worked in Turin, Berlin, and Paris. He contributed to variational calculus, mathematical analysis, and mechanics. His *Mécanique Analytique* is still a valuable text, although he refused to use pictures. Lagrange's formulation of mechanics is named after him, as are the three equilibrium solutions (Lagrange points) that he found in the three-body problem.

Edmond N. Laguerre (1834–1886), French mathematician who entered the École Polytechnique in 1852. Although physically not very strong, he began a career in the army in 1854, returning to the École Polytechnique in 1864. His work was in analysis and geometry.

Pierre-Simon Laplace (1749–1827), French mathematician and astronomer with fundamental contributions in probability, potential theory (Laplace equation), and celestial mechanics. His five volumes on celestial mechanics contain many long (formal) expansions for the three-body problem. His social and political aptitude

Fig. 13.5 Leopold
Kronecker, mathematician

helped him to survive the French revolution, the regime of Napoleon, and the Bourbon restoration, and he ended his life with the rank of a marquis.

Gustave Le Bon (1841–1931); see Chapter 6 and Figure 6.1.

Édouard Le Roy (1870–1954), French philosopher and mathematician, student of Bergson. His idea was to combine conventionalism with his Catholic faith. The Vatican rejected his "modernist" works.

Urbain J.J. Le Verrier (1811–1877), French astronomer, director of the Paris Observatory. He predicted the existence of the planet Neptune on the basis of orbital irregularities of Uranus. He caused his colleagues considerable trouble.

Anders Lindstedt (1854–1939), Swedish mathematician who worked in Sweden and Russia. He formulated an approximation method for nonlinear differential equations that was taken up by Henri Poincaré and now bears both their names. As a professor at Stockholm University, he became interested around 1900 in actuarial science. From 1909 on, he worked full time on insurance problems, advising private companies and the Swedish government.

Gabriel Lippmann (1845–1921), born in Luxembourg, entered the École Normale Supérieure in 1868. Doctorate in Heidelberg, 1874; appointed to a physics chair at the Sorbonne in 1878, received the Nobel Prize in physics in 1908. Most famous student: Couette.

Hendrik A. Lorentz (1853–1928), Dutch physicist who studied in Leiden and was appointed to a chair at Leiden university in 1878. Most important work was on

Fig. 13.6 Colonel Amédée
Mannheim, mathematician

electromagnetic theory, theory of the electron, and relativity. He was awarded the
Nobel Prize in physics in 1908. See Figure 11.4.

Aleksandr M. Lyapunov (1857–1919), Russian mathematician who studied in St.
Petersburg. Doctorate in Moscow 1892; 1893–1902 occupied a chair at Kharkov
University, 1902–1917 at the University of St. Petersburg. His achievements are on
the stability of motion and applications in mechanics.

Colin Maclaurin (1698–1746), Scottish mathematician, studied in Glasgow. Most
important work on series expansions and equilibria of rotating fluid masses.

Victor M. Amédée Mannheim (1831–1906), entered the École Polytechnique in
1848, went into the army, and was promoted to colonel. Returned to the École
Polytechnique in 1859, appointed to a chair in descriptive geometry in 1864. Known
as the inventor of the modern slide rule. See Figure 13.6.

Frédéric Masson (1847–1923), French historian, member of the Académie Française
from 1903, seat 17.

James Clerk Maxwell (1831–1879), Scottish mathematical physicist, born in Ed-
inburgh. His equations form the foundations of electromagnetic theory. He also
demonstrated that the rings of Saturn cannot be fluid or solid. He published as a
student and held positions at Aberdeen, King's College (London), and Cambridge.

Hermann Minkowski (1864–1909), German-Jewish mathematician, born in Lithua-
nia. He studied in Berlin and Königsberg, published in number theory, and showed
that the theory of special relativity of Lorentz, Poincaré, and Einstein could be cast
in the framework of four-dimensional non-Euclidean geometry. From 1902 he held
a position in Göttingen as a colleague of Hilbert.

Gösta Mittag-Leffler (1846–1927); see Section 4.4 and Figure 4.4.

Fig. 13.7 Jürgen K. Moser, mathematician

Jürgen K. Moser (1928–1999), born in Königsberg and as a refugee from East Germany, studied in Göttingen, where he completed his doctorate under the direction of F. Rellich. In 1953, he moved to the United States, became an American citizen, and was appointed director of the Courant Institute of Mathematical Sciences (1967–1980). The famous Moser twist theorem, which is basic to the KAM theorem, was published in 1962. From 1980 to 1995, Moser was director of the ETH Zurich, in Switzerland. See Figure 13.7.

Forest Ray Moulton (1872–1952), came from a family of American pioneers and was the first in his town to attend college. He studied astronomy at the University of Chicago and taught there until he left for public service in 1926. He became known for his books and papers on celestial mechanics and differential equations, also for the planetesimal hypothesis, developed together with T.C. Chamberlin.

Simon Newcomb (1835–1909) was a Canadian–American mathematician and astronomer. He studied at Harvard University and was attached to the US Naval Observatory and the Nautical Almanac Office. He held several professorships, for instance at John Hopkins University. His main interest was in celestial mechanics, in particular the calculation of the positions of planets and the Moon for publication in ephemerides.

Paul Painlevé (1863–1933), French mathematician, doctorate in Paris 1883; after Lille, appointment at the Sorbonne in 1892. In 1906, he began a political career, leaving science in 1910 for politics and serving as prime minister during two periods.

Giuseppe Peano (1858–1933), born in Spinetta (Piemonte), Italy, studied and worked in Turin. He became famous for his existence theorem for ODEs and his axiomatics of the natural numbers.

Lars Edvard Phragmén (1863–1937), Swedish mathematician, studied in Uppsala and Stockholm, where he soon became an editorial assistant of Mittag-Leffler for the *Acta Mathematica*. In 1892, he became a professor at the University of Stockholm, and in 1904, he left the university for insurance work, first in a public position, later private. See Figure 5.2.

Charles Émile Picard (1856–1941), French mathematician who was educated at the École Normale Supérieure in Paris. He lectured at Toulouse University and later in Paris. His mathematical achievements were in analysis (special functions and his successive approximation–contraction method for nonlinear differential equations), algebraic geometry, and mechanics. After World War I, he was one of the leaders of a boycott of German scientists.

Max Planck (1858–1947), German physicist who studied in Munich and Berlin; he became one of the founders of quantum mechanics and was awarded the Nobel Prize in physics in 1918. As a leading physics professor in Berlin, he experienced difficulties during the Nazi period.

Siméon-Denis Poisson (1781–1840) was a French mathematician. He studied at the École Polytechnique in Paris, where Lagrange and Laplace lectured. He obtained many results in applied mathematics, probability, mathematical physics, and celestial mechanics. In 1806, he succeeded Fourier as a professor in Paris, and in 1827, Laplace. He was made a baron but declined to use the title.

René François Armand (Sully) Prudhomme (1839–1907), French poet, received the first Nobel Prize for literature in 1901 and became a member of the Académie Française in 1881; his successor at the Académie was Henri Poincaré.

Victor Puiseux (1820–1883) was a French mathematician and astronomer. Educated at the École Normale Supérieure, doctorate in astronomy and mechanics 1841. He held various university positions, worked on elliptic functions and series expansions, succeeded Cauchy in 1857 to the Sorbonne chair of mathematical astronomy.

Bernhard Riemann (1826–1866), studied mathematics in Göttingen and completed his habilitation under the supervision of Gauss. He made fundamental contributions to complex analysis and differential geometry. See Figure 13.8.

Édouard Roche (1820–1883), French mathematician and astronomer who studied and worked at Montpellier. Of his work in celestial mechanics and cosmogony, most famous is the Roche limit, which gives the distance at which a moon of a planet is destroyed by tidal forces.

Eugène Rouché (1832–1910) was a productive mathematician who taught at schools of higher education in Paris. Some of his theorems and also textbooks on geometry and analysis became well known. Together with Hermite and Poincaré, he edited Laguerre's collected works.

Fig. 13.8 Bernhard
Riemann, mathematician

Hermann Schwarz (1824–1907) moved from chemistry to mathematics under the influence of Kummer and Weierstrass. He obtained a position in Göttingen, and in 1892, a chair in Berlin. His contributions are concerned with complex analysis and variational problems.

Karl Schwarzschild (1873–1916) was born in Frankfurt-am-Main, Germany, into a highly cultured German-Jewish family. As a schoolboy, he published two papers on celestial mechanics, studied in Strasbourg, and received his doctorate in Munich on Poincaré's theory of rotating fluid masses. In 1900, he suggested in a paper that space is non-Euclidean; later, he gave the first exact solution of Einstein's equations of general relativity. From 1901 he was a professor in Göttingen, from 1909 in Potsdam.

Thomas J.J. See (1866–1962), astronomer, born near Montgomery City, Missouri, USA, and graduated from the University of Missouri. Completed his doctorate in Berlin, but on returning to the United States had a problematic career. He specialized in double stars and cosmogony.

Carl L. Siegel (1896–1981) was born in Berlin, Germany, and studied in Göttingen. He became an important mathematician with a main interest in number theory and celestial mechanics. As a convinced opponent of the Nazi regime, he emigrated to the United States in 1940, where he worked at the Institute of Advanced Study, in Princeton. In 1951 he accepted an appointment as professor in Göttingen, and in 1978 he was awarded the Wolf prize. His most famous student was Jürgen K. Moser.

Willem de Sitter (1872–1934) was born in Groningen, The Netherlands. He became a professor of astronomy at Leiden University. His scientific activity was in celestial mechanics, especially the satellites of Jupiter, and cosmology (the Einstein–de Sitter model).

Arnold J.W. Sommerfeld (1868–1951) was a German mathematical physicist who studied in Göttingen and did his habilitation under Felix Klein. His first chair was in Aachen (1900), and in 1906 he moved to Munich. He had many brilliant students, including Ludwig Hopf, Werner Heisenberg, Wolfgang Pauli, and Hans Bethe.

Thomas Jan Stieltjes (1856–1894), born in Zwolle, The Netherlands, failed his examination at the Technical University, Delft, and became an assistant at Leiden Observatory. He simplified Tisserand's calculations and exchanged letters with Hermite; this correspondence continued and contains 432 letters. In 1884, he was awarded an honorary doctorate at Leiden. He was invited by Darboux and Hermite to Paris, where in 1886, he defended his doctoral dissertation on semiconvergent (asymptotic) series. Also in 1886 he became Maître de Conférences in Toulouse.

William Thomson (Lord Kelvin) (1824–1907), born in Belfast, Ireland, British mathematical physicist and engineer–inventor. Together with P.G. Tait, he wrote the very influential *Treatise on Natural Philosophy* (1867).

Felix Tisserand (1845–1896) was a French astronomer, educated at the École Normale Supérieure in Paris. He was director of the observatory in Toulouse and from 1892 of the Paris Observatory. He extended Delaunay's studies of the three-body problem and published four volumes on celestial mechanics [Tisserand 1889]. These books are considered an update of Laplace's celestial mechanics and became a standard reference in the field.

Karl Theodor Wilhelm Weierstrass (1815–1897) had a difficult start in mathematics. He worked first as a secondary-school teacher but eventually was appointed to a chair in Berlin. His rigorous style of mathematical analysis became the standard in Germany and in many other places. He influenced many students who became important later. He recognized the quality of the mathematics of Sonya Kovalevskaya.

Wilhelm Wien (1864–1928), German physicist, whose doctoral dissertation was supervised by Helmholtz. He was awarded the Nobel Prize in physics in 1911.

Pieter Zeeman (1865–1943), born in Zeeland, The Netherlands, began his physics study in Leiden under the supervision of Kamerlingh Onnes and Lorentz. He was awarded the Nobel Prize in physics in 1902 (together with Lorentz) for his work on the magnetic splitting of spectral lines. He succeeded J.D. van der Waals at Amsterdam University in 1908.

Ernst F.F. Zermelo (1871–1953) was a German mathematician, educated in Berlin; he was professor in Zurich, later in Freiburg im Breisgau, from which he resigned in 1935 because of his disapproval of the Nazi regime. His activity concerned the foundations of mathematics, in particular axiomatic set theory. He objected to Boltzmann's statistical mechanics because of his interpretation of the Poincaré recurrence theorem.

References

[Appell 1880] Paul Appell. "Sur une classe de polynomes." *Ann. Sci. École Norm. Sup.* 9 (1880), 119–144.

[Appell 1891] Paul Appell. "Sur les lois de forces centrales faisant décrire à leur point d'application une conique quelles soient les conditions initiales." *Am. J. Math.* 13 (1891), 151–158.

[Appell 1900] Paul Appell. "Sur une forme générale des équations de la dynamique et sur le principe de Gauss." *J. Reine Angew. Math.* 122 (1900), 205–208.

[Appell 1921] Paul Appell. *Traité de Mécanique Rationelle*, 4 vols., Paris: Gauthier-Villars, 1921–1941.

[Appell 1925a] Paul Appell. *Henri Poincaré*, Paris: Librairie Plon, 1925.

[Appell 1925b] Paul Appell. "Notice sur les travaux scientifiques." *Acta Mathematica* 45 (1925), 161–285.

[Arnold 1978] V.I. Arnold, *Mathematical Methods of Classical Mechanics*. New York: Springer-Verlag, 1978.

[Arnold 1983] V.I. Arnold. *Geometrical Methods in the Theory of Ordinary Differential Equations*. New York: Springer-Verlag, 1983.

[Arnold 1993] V.I. Arnold, editor. *Dynamical Systems VIII*. In *Encyclopaedia of Mathematical Sciences*. Berlin: Springer, 1993.

[Arnold et al. 1988] V.I. Arnold, V.V. Kozlov, and A.I. Neihstadt, editors. "Mathematical aspects of classical and celestial mechanics." In *Dynamical Systems III*, edited by V.I. Arnold, *Encyclopaedia of Mathematical Sciences*. Berlin: Springer, 1988.

[Barrès et al. 1954] Philippe Barrès et al. *Le Livre des centenaires*, Nancy, 1954; recollections at 150 years of the lycée of Nancy, now called "Lycée Henri Poincaré."

[Barrow-Green 1997] June Barrow-Green. *Poincaré and the Three Body Problem*, History of Mathematics 11. Providence: American Mathematical Society, and London: London Mathematical Society, 1997.

[Bellivier 1956] André Bellivier, *Henri Poincaré, ou la vocation souveraine*. Paris: Gallimard, 1956.

[Benettin et al. 1982] G. Benettin, G. Ferrari, L. Galgani, and A. Giorgilli. "An extension of the Poincaré–Fermi theorem on the nonexistence of invariant manifolds in nearly integrable Hamiltonian systems." *Il Nuovo Cimento* 72B (1982), 137–148.

[Birkhoff 1913] G.D. Birkhoff. "Proof of Poincaré's geometric theorem." *Trans. AMS* 14 (1913), 14–22.

[Birkhoff 1927] G.D. Birkhoff. "On the periodic motions of dynamical systems." *Acta Mathematica* 50 (1927), 359–379.

[Birnbaum 1994] P. Birnbaum. *L'affaire Dreyfus, la république en péril*. Paris: Gallimard, 1994.

F. Verhulst, *Henri Poincaré: Impatient Genius*, DOI 10.1007/978-1-4614-2407-9,
© Springer Science+Business Media New York 2012

[Borel 1914] Émile Borel. *Introduction géométrique à quelques théories physiques*. Paris: Gauthier-Villars, 1914.

[Boutroux 1912] Aline Boutroux. *Vingt ans de ma vie, simple vérité*. Nancy: Archives B Centre d'Études et de Recherches Henri Poincaré, 1912.

[Boutroux 1921] P. Boutroux. "Lettre de Pierre Boutroux à M. Mittag-Leffler." *Acta Math.* 38 (1921), 197–201.

[Briot and Bouquet 1856] C. Briot and T. Bouquet. "Recherches sur les propriétés des fonctions définies par les équations différentielles." *J. de l'École Polytechnique*, Cahier 21 (1856), 133–198.

[Broer 2004] H.W. Broer. "KAM theory: the legacy of Kolmogorov's 1954 paper." *Bull. AMS* 41 (2004), 507–521.

[Brouwer 1908] L.E.J. Brouwer. "Letter to D.J. Korteweg." Brouwer Archive DJK 39, 1908.

[Browder 1983] Felix E. Browder, editor. *Mathematical Heritage of Henri Poincaré*. Proc. Symposia on Pure Mathematics, 39. Providence: AMS, 1983.

[Bruns 1888] H. Bruns. "Über die Integrale des Vielkörper-Problems." *Acta Mathematica* 11 (1888), 25–96.

[Brush 1996] Stephen G. Brush. *Fruitful Encounters: The Origin of the Solar System and of the Moon from Chamberlin to Apollo*. Cambridge: Cambridge University Press, 1996.

[Cartan 1922] Élie Cartan. *Leçons sur les invariants intégraux*. Paris: Hermann, 1922.

[Cartan 1974] Élie Cartan. *Notice sur les travaux scientifiques, suivi de Le parallélisme absolu et la théorie unitaire du champ*. Paris: Gauthier-Villars, 1974.

[Charpentier et al. 2010] E. Charpentier, E. Ghys, and A. Lesne, editors. *The Scientific Legacy of Henri Poincaré*, History of Mathematics 36. Providence: AMS, and London: London Math. Society, 2010 (translation by Joshua Bowman from original French edition by Editions Belin, Paris, 2006).

[Chasles 1880] Michel Chasles, *Discours d'Inauguration du Cours de Géométrie de Faculté des Sciences, Paris*, 22 Decembre 1846. Also included in *Traité de Géométrie Supérieure*, 2e ed. Paris: Gauthier-Villars, 1880.

[Chebyshev 1907] P.L. Tchebychef (Chebyshev). *Oeuvres* vol. 2, edited by A. Markoff and N. Sonin. St. Petersburg, 1907.

[Chicone 1999] C. Chicone, *Ordinary Differential Equations with Applications*, Texts in Applied Mathematics 34. New York: Springer, 1999.

[Counselman 1973] Charles C. Counselman III. "Outcomes of tidal evolution." *Ap. J.* 180 (1973), 307–314.

[Darboux 1913] Gaston Darboux. "Éloge historique d'Henri Poincaré." In [Poincaré 1916], vol. 2, pp. VII–LXXI, presented as a lecture on December 5, 1913.

[de Sitter 1907] Willem de Sitter. "On the libration of the three inner large satellites of Jupiter." *Publ. Astr. Lab. Groningen* 17 (1907), 1–119, 1907 (see also *Ann. Sterrewacht* Leiden vol. 12, 1925).

[Einstein 1950] Albert Einstein. "The theory of relativity." In *Out of My Later Years*. New York: Philosophical Library, 1950.

[Eymar 1996] P. Eymar. "Comment Hilbert et Poincaré rédigeaient les mathématiques." *Philosophia Scientia*, Nancy 1 (1996), 19–26.

[Fermi 1923] E. Fermi. "Generalizzazione del teorema di Poincaré sopra la non esistenza di integrali uniformi di un sistema di equazioni canoniche normali." *Il Nuovo Cimento* 26 (1923), 105–115.

[Freudenthal 1954] H. Freudenthal. "Poincaré et les fonctions automorphes." in [Poincaré 1916] vol. 11, pp. 212–219, 1954.

[Fuchs 1880] L.I. Fuchs. "Über eine Klasse von Functionen meherer Variabeln, welche durch Umkehrung der Integrale von Lösungen der linearen Differentialgeleichungen mit rationalen Coefficienten entstehen." *J. Reine Angew. Math.* 89 (1880), 151–169.

[Ghys 2010] E. Ghys. "Variations on Poincaré's recurrence theorem." In *The Scientific Legacy of Poincaré*, edited by E. Charpentier, E. Ghys, and A. Lesne, History of mathematics 36, pp. 193–206. Providence: AMS, 2010.

[Gray and Walters 1997] Jeremy J. Gray and Scott A. Walters, editors. *Henri Poincaré, Three Supplementary Essays on the Discovery of Fuchsian Functions.* Berlin: Akademie Verlag, and Paris: Albert Blanchard, 1997.

[Hadamard 1990] J. Hadamard. *The Theory of Invention in the Mathematical Field.* New York: Dover, 1990.

[Hadamard 2000] J. Hadamard. *Non-Euclidean Geometry in the Theory of Automorphic Functions,* edited by Jeremy J. Gray and Abe Shenitzer, History of Mathematics 17. Providence: AMS, and London: London Math. Soc., 2000.

[Hale 1969] J.K. Hale. *Ordinary Differential Equations.* New York: Wiley-Interscience, 1969.

[Heinzmann 2010] G. Heinzmann. "Henri Poincaré and his thoughts on the philosophy of science." In *The Scientific Legacy of Poincaré,* edited by E. Charpentier, E. Ghys, and A. Lesne, History of mathematics 36, pp. 373–391. Providence: AMS, 2010.

[Hermite 1985] Letter from C. Hermite to G. Mittag-Leffler (1888). In *Cahiers de Séminaire d'Histoire des Mathématiques* 6 (1985), 146.

[Hilbert 1899] D. Hilbert. *Grundlagen der Geometrie.* Teubner: Leipzig, 1899.

[Klein 1894] F. Klein. "Ueber lineare Differentialgleichungen der zweiten Ordnung." Vorlesung, gehalten im Sommersemester 1894, ausgearbeitet von E. Ritter, Göttingen, 1894.

[Kirillov and Verhulst 2010] O.N. Kirillov and F. Verhulst. "Paradoxes of dissipation-induced destabilization or who opened Whitney's umbrella." *ZAMM* 90 (2010), 462–488.

[Klein 1924] F. Klein. "Correspondence d'Henri Poincaré et de Felix Klein," edited by N.E. Nörlund. *Acta Mathematica* 39 (1924), 94–132.

[Kozlov 1996] V.V. Kozlov. *Symmetries, Topology and Resonances in Hamiltonian Mechanics,* Ergebnisse der Mathematik und ihre Grenzgebiete 31. New York: Springer, 1996.

[Le Bellac 2010] M. Le Bellac. "The Poincaré group." In *The Scientific Legacy of Poincaré,* edited by E. Charpentier, E. Ghys and A. Lesne, History of Mathematics 36, pp. 329–350. Providence: AMS, 2010.

[Lebon 1912] Ernest Lebon, *Henri Poincaré, biographie, bibliographie analytique des écrits,* 2nd ed. Paris: Gauthier-Villars, 1912.

[Lebovitz 1998] N.R. Lebovitz. "The mathematical development of the classical ellipsoids." *Int. J. Eng. Science* 36 (1998), 1407–1420.

[Lorentz 1904] H.A. Lorentz. "Electromagnetic phenomena in a system moving with any velocity smaller than that of light." In *Proc. Royal Academy of Sciences, Amsterdam,* May 27, 1904. Reprinted in collected papers vol. 5, pp. 172–197. The Hague: Martinus Nijhoff, 1939.

[Lorentz 1914] H.A. Lorentz. "Deux mémoires de Henri Poincaré dans la physique mathématique." *Acta Mathematica* 38 (1914), 293–308 (reprinted in collected papers vol. 8, pp. 258–273. The Hague: Matinus Nijhoff, 1934).

[Lorentz 1915] H.A. Lorentz. "De lichtaether en het relativiteitsbeginsel" (the light-ether and the principle of relativity). Lecture for the Royal Academy of Sciences, Amsterdam, April 24, 1915; Reprinted in collected papers vol. 9, pp. 233–243. The Hague: Martinus Nijhoff, 1939.

[Lorentz 1928] H.A. Lorentz. "Conference on the Michelson–Morley Experiment." *Astrophysical Journal* 68 (1928), 345–351.

[Mannheim 1909] Amédée Mannheim. "L'invention mathématique, note posthume." *L'Enseignement Mathématique* 11 (1909), 161–167.

[Mawhin 2010] J. Mawhin. "Henri Poincaré and the partial differential equations of mathematical physics." In *The Scientific Legacy of Poincaré,* edited by E. Charpentier, E. Ghys, and A. Lesne, History of mathematics 36, pp. 373–391. Providence: AMS, 2010.

[Mooij 1966] J.J.A. Mooij, *Philosophie des mathématiques de Henri Poincaré.* Paris: Gauthier-Villars, 1966.

[Nabonnand 1999] Philippe Nabonnand. "The Poincaré–Mittag-Leffler relationship." *The Mathematical Intelligencer* 21 (1999), 58–64.

[Novikov 2004] S. Novikov. "Henri Poincaré and XXth century topology." In *Solvay Workshops and Symposia,* vol. 2, Symposium Henri Poincaré, edited by P. Gaspard, M. Henneaux, and F. Lambert, pp. 17–24, 2004.

[Osgood 1938] W.F. Osgood. *Functions of a Complex Variable.* New York: G.E. Stechert, 1938.

[Picard 1891] Émile Picard. *Traité d'Analyse*, 3 vols. Paris: Gauthier-Villars, 1891, 1893, 1896.

[Poincaré 1878] Henri Poincaré. "Note sur les propriétés des fonctions définies par les équations différentielles." *J. de l'École Polytechnique*, Cahier 45 (1878), 13–26; also in [Poincaré 1916] vol. 1, pp. XXXVI–XLVIII.

[Poincaré 1881] Henri Poincaré. "Mémoire sur les courbes définies par une équation différentielle." *J. de Mathématiques*, 3e série vol. 7 (1881), 375–422, vol. 8 (1882), pp. 251–296.

[Poincaré 1882] Henri Poincaré. "Sur les fonctions uniformes qui se reproduisent par des substitutions linéaires." *Math. Annalen* 19 (1882), 553–564; also in [Poincaré 1916] vol. 2, pp. 92–105.

[Poincaré 1886] Henri Poincaré. "Sur les intégrales irrégulières des équations linéraires." *Acta Mathematica* 8 (1886), 295–344.

[Poincaré 1890a] Henri Poincaré. *Lecture notes:*

1. *Théorie mathématique de la lumière*, vol. 1, 408 pp., Georges Carré et C. Naud, Paris, 1889.
   Théorie mathématique de la lumière, vol. 2, *Nouvelles études sur la diffraction, Théorie de la dispersion de Helmholtz*, 310 pp., Georges Carré et C. Naud, Paris, 1892, edited by J. Blondin, M. Lamotte, and D. Hurmuzescu.

2. *Électricité et optique*, vol. 1, *Les théories de Maxwell et la théorie électromagnetique de la lumière*, 314 pp., Georges Carré, Paris, 1890, edited. J. Blondin.
   Électricité et optique, vol. 2, *Les théories de Helmholtz et les expériences de Hertz*, XI+262 pp., Georges Carré, Paris, 1891, edited by B. Brunhes. Second edition 658 pp., Gauthier-Villars, Paris, 1901, edited by J. Blondin and E. Neculcea.

3. *Thermodynamique*, 432 pp., Georges Carré et C. Naud, Paris, 1892, edited by J. Blondin. Second edition Gauthier-Villars, Paris, 1908, edited by J. Blondin.

4. *Leçons sur la théorie de l'elasticité*, 210 pp., Georges Carré et C. Naud, Paris, 1892, edited by É. Borel and J. Drach.

5. *Théorie des tourbillons*, 212 pp., Georges Carré, Paris, 1893, edited by M. Lamotte.

6. *Les oscillations électriques*, 343 pp., Georges Carré et C. Naud, Paris, 1894, edited by C. Maurain.

7. *Capillarité*, 189 pp., Georges Carré, Paris, 1895, edited by J. Blondin.

8. *Théorie analytique de la propagation de la chaleur*, 316 pp., Georges Carré, Paris, 1895, edited by L. Rouyer and R. Baire.

9. *Calcul des probabilités*, 275 pp., Gauthier-Villars, Paris, 1896, edited by A. Quiquet.

10. *Théorie du potentiel Newtonien*, 366 pp., Gauthier-Villars, Paris, 1899, edited by E. le Roy and G. Vincent. 2d ed. Gauthier-Villars, Paris, 1912.

11. *Cinématique et mécanismes, potentiel et mécanique des fluides*, 385 pp., Georges Carré et C. Naud, Paris, 1899, edited by A. Guillet.

12. *Figures d'équilibre d'une masse fluide*, 211 pp. C. Naud, Paris, 1902, edited by L. Dreyfus.

13. *Leçons de mécanique céleste*, Gauthier-Villars, Paris.
    Vol. 1, *Théorie générale des perturbations planetaires*, 368 pp. 1905.
    Vol. 2, part 1: *Développement de la fonction perturbatrice*, 168 pp. 1907
    Vol. 2, part 2: *Théorie de la lune*, 138 pp. 1909.
    Vol. 3, *Théorie des marées*, 469 pp. 1910, third volume edited by E. Fichot.

14. *Leçons sur les hypothèses cosmogoniques*, 294 pp., A. Hermann et fils, Paris, 1911, edited by H. Vergne. Second edition: A. Hermann et fils, Paris, 1913 with "Notice sur Henri Poincaré" by Ernest Lebon (43 pp.).

[Poincaré 1890b] Henri Poincaré. "Sur le problème des trois corps et les équations de la dynamique." *Acta Mathematica* 13 (1890), 1–270; also in [Poincaré 1916] vol. 7, pp. 262–479.

[Poincaré 1890c] Henri Poincaré. "Sur les équations aux dérivées partielles de la physique mathématique." *American J. Math.* 12 (1890), 211–294; also in [Poincaré 1916] vol. 9, pp. 28–113.

[Poincaré 1892] Henri Poincaré. *Les Méthodes Nouvelles de la Mécanique Céleste*, 3 vols. Paris: Gauthier-Villars, 1892, 1893, 1899.

[Poincaré 1894a] Henri Poincaré. "Sur l'équation des vibrations d'une membrane." *C.R.A.S. Paris* 118:12 (1894) 447–451; also in [Poincaré 1916] vol. 9, pp. 119–122.

[Poincaré 1894b] Henri Poincaré. "Sur les équations de la physique mathématique." *Rend. Circolo Mat. Palermo* 8 (1894), 57–155; also in [Poincaré 1916] vol. 9, pp. 123–196.

[Poincaré 1895a] Henri Poincaré. "À propos de la théorie de M. Larmor." *L'Éclairage Électrique* 3, 5 (1895); also in [Poincaré 1916], vol. 9, pp. 369–426.

[Poincaré 1895b] Henri Poincaré. "La théorie de Lorentz et la principe de réaction." *Archives Néerlandaises des Sciences Exactes et Naturelles*, 2e Série 5 (1895), 252–278; also in [Poincaré 1916], vol. 9, pp. 464–488.

[Poincaré 1902] Henri Poincaré. *La Science et l'Hypothèse*. Paris: Flammarion, 1902.

[Poincaré 1905a] Henri Poincaré. "Sur la dynamique de l'électron. *Comptes Rendus de l'Académie des Sciences* 140 (1905), 1504–1508; also in [Poincaré 1916], vol. 9, pp. 489–493.

[Poincaré 1905b] Henri Poincaré. *La Valeur de la Science*. Paris: Flammarion, 1905.

[Poincaré 1906] Henri Poincaré. "Sur la dynamique de l'électron." *Rendiconti del Circolo Matematico di Palermo* 21 (1906), 129–176; also in [Poincaré 1916], vol. 9, pp. 494–550.

[Poincaré 1908a] Henri Poincaré. *Science et Méthode*. Paris: Flammarion, 1908.

[Poincaré 1908b] Henri Poincaré. "Conférence sur la télégraphique sans fil." *Revue d'électricité* (27 Décembre 1908), 387–393.

[Poincaré 1909] Henri Poincaré. *Sechs Vorträge aus der reinen Mathematik und mathematischen Physik*, Mathematische Vorlesungen an der Universität Göttingen: IV, auf Einladung der Wolfskehl-Kommission der Königlichen Gesellschaft der Wissenschaften gehalten zu Göttingen vom 22–28 April 1909. Leipzig and Berlin: B.G. Teubner, 1910.

[Poincaré 1910] Henri Poincaré. *Savants et Écrivains*. Paris: Flammarion, 1910.

[Poincaré 1911] Henri Poincaré. *Les sciences et les humanités*. Paris: A. Fayard, 1911.

[Poincaré 1912a] Henri Poincaré. "Les rapports de la matière et de l'éther." *J. de Physique théorique et appliquée* 2 (1912), 347–360; lecture at a conference organized by the Société française de Physique, see also [Poincaré 1916] vol. 9, pp. 669–682.

[Poincaré 1912b] Henri Poincaré, *Sur un théorème de géometrie*, Rend. Circolo Mat. Palermo 33, pp. 375-407, 1912; also in [Poincaré 1916], vol. 6, pp. 499–538.

[Poincaré 1913] Henri Poincaré, *Dernières Pensées*, Flammarion, Paris, 1913.

[Poincaré 1916] Henri Poincaré, *Oeuvres de Henri Poincaré publiées sous les auspices de l'Académie des Sciences*, vols. 1–12, Gauthier-Villars, Paris, 1916–1954.

[Poincaré 1985] Henri Poincaré, *Papers on Fuchsian Functions*, translated and introduced by John Stillwell, Springer, 1985.

[Poincaré 2002] Henri Poincaré, *Scientific Opportunism: An Anthology*, Compiled by Louis Rougier, ed. by Laurant Rollet, Birkhäuser (Publications des Archives Henri-Poincaré), Basel, 2002.

[Poincaré 2010] Henri Poincaré, *Papers on topology, Analysis Situs and its five supplements*, AMS and London Math. Soc., History of mathematics vol. 37, 2010.

[Poincaré 1999] Henri Poincaré and Gösta Mittag-Leffler. *La correspondance entre Henri Poincaré et Gösta Mittag-Leffler, avec en annexes les lettres échangées par Poincaré avec Fredholm, Gyldén et Phragmén*; presentée et annotée par Philippe Nabonnand. Basel: Birkhäuser, 1999.

[Poincaré 2012] Henri Poincaré. Correspondence Archives Henri Poincaré, Université de Nancy.

[Russell 1914] Bertrand Russell. Preface to an English translation of *Science et Méthode* by Francis Maitland. London: Thomas Nelson and Sons (no date but after 1913).

[Sanders et al. 2007] J.A. Sanders, F. Verhulst, and J. Murdock. *Averaging Methods in Nonlinear Dynamical Systems*, second revised edition, Applied Math. Sciences 59. New York: Springer, 2007.

[Sanzo 1996] U. Sanzo. "Contre l'ontologie: Poincaré et les hypothèses scientifiques." *Philosophia Scientia Nancy* 1 (1996), 27–43.

[Schwarzschild 1913] K. Schwarzschild. "Review of Poincaré's 'Leçons sur les hypothèses cosmogoniques.'" *Ap. J.* 37 (1913), 294–298.

[See 1912] T.J.J. See. "Review of Poincaré's lectures on cosmogony." *Review of Popular Astronomy* 20 (1912), 13–14.

[Siegel and Moser 1971] C.L. Siegel and J.K. Moser. *Lectures on Celestial Mechanics*, Grundlehren der mathematischen Wissenschaften 187. New York: Springer, 1971.

[Smale 1967] S. Smale. "Differentiable dynamical systems." *Bull. AMS* 73 (1967) 747–817.

[Steckline 1983] V.S. Steckline. "Zermelo, Boltzmann and the recurrence paradox." *Am. J. Physics* 51(1983), 894–897.

[Stieltjes 1886] T.-J. Stieltjes. "Recherches sur quelques séries semi-convergentes." *Ann. Scientifiques de École Normale Supérieure* 3 (1886), 201–258.

[Thom 1987] René Thom. "La philosophie des sciences d'Henri Poincaré." *Cahiers d'Histoire et de Philosophie des Sciences* 23, 1987.

[Thomson and Tait 1883] W. Thomson and P.G. Tait. *Treatise on Natural Philosophy.* Cambridge: Cambridge University Press, 1883.

[Tisserand 1889] Félix Tisserand. *Traité de mécanique céleste*, 4 vols. Paris: Gauthier-Villars, 1889–1896.

[Toulouse 1910] E. Toulouse. *Henri Poincaré.* Paris: Flammarion, 1910.

[Van Dalen 1999] D. Van Dalen. *Mystic, Geometer and Intuitionist: The Life of L.E.J. Brouwer*, vol. 1, *The Dawning Revolution.* Oxford: Clarendon Press, 1999.

[Van der Aa and De Winkel 1994] E. Van der Aa and M. De Winkel. "Hamiltonian systems in 1 : 2 : ω (ω = 5 or 6) resonance." *Int. J. Nonlin. Mech.* 29 (1994) 261–270.

[Verhulst 2000] Ferdinand Verhulst. *Nonlinear Differential Equations and Dynamical Systems*, second revised edition. New York: Springer, 2000.

[Verhulst 2005] Ferdinand Verhulst. *Methods and Applications of Singular Perturbations.* New York: Springer, 2005.

[Volterra et al. 1914] Vito Volterra, Jacques Hadamard, Paul Langevin, and Pierre Boutroux. *Henri Poincaré, l'oeuvre scientifique, l'oeuvre philosophique.* Paris: Librairie Félix Alcan, 1914.

Index